LEARNING TO LIVE WITH DATAFICATION

As digital technologies play a key role across all aspects of our societies and in everyday life, teaching students about data is becoming increasingly important in schools and universities around the world. Bringing together international case studies of innovative responses to datafication, this book sets an agenda for how teachers, students and policy makers can best understand what kind of educational intervention works and why.

Learning to Live with Datafication is unique in its focus on educational responses to datafication as well as critical analysis. Through case studies grounded in empirical research and practice, the book explores the dimensions of datafication from diverse perspectives that bring in a range of cultural aspects. It examines how educators conceptualise the social implications of datafication and what is at stake for learners and citizens as educational institutions try to define what datafication will mean for the next generation.

Written by international leaders in this emerging field, this book will be of interest to teacher educators, researchers and post graduate students in education who have an interest in datafication and data literacies.

Luci Pangrazio is a senior lecturer and Alfred Deakin postdoctoral research fellow at Deakin University, Melbourne, Australia. Her research studies personal data and privacy, the politics of digital platforms and young people's critical understandings of digital media. She is currently researching methods for visualising and understanding digital data for educational purposes. Her book *Young People's Literacies in the Digital Age: Continuities, Conflicts and Contradictions* was published in 2019 by Routledge.

Julian Sefton-Green is a Professor of New Media Education at Deakin University, Melbourne, Australia. He has worked as an independent scholar and has held positions at the Department of Media & Communication, London School of Economics & Political Science, and at the University of Oslo. He has researched and written widely on many aspects of media education, new technologies, creativity, digital cultures and informal learning and has authored, co-authored or edited 18 books and has spoken at over 50 conferences in over 20 countries.

Both editors are chief investigators at the *Australian Research Council Centre of Excellence for the Digital Child.*

LEARNING TO LIVE WITH DATAFICATION

Educational Case Studies and Initiatives from Across the World

Edited by
Luci Pangrazio and Julian Sefton-Green

Routledge
Taylor & Francis Group
LONDON AND NEW YORK

Cover image by Anne Scott Wilson, Coat hanger, 2016,
http://www.annescottwilson.com

First published 2022
by Routledge
2 Park Square, Milton Park, Abingdon, Oxon OX14 4RN

and by Routledge
605 Third Avenue, New York, NY 10158

Routledge is an imprint of the Taylor & Francis Group, an informa business

British Library Cataloguing-in-Publication Data
A catalogue record for this book is available from the British Library

Library of Congress Cataloging-in-Publication Data
Names: Pangrazio, Luci, editor. | Sefton-Green, Julian, editor.
Title: Learning to live with datafication : educational case studies and initiatives from across the world / Edited by Luci Pangrazio and Julian Sefton-Green.
Description: Abingdon, Oxon ; New York, NY : Routledge, 2022.
| Includes bibliographical references and index.
| Identifiers: LCCN 2021042727 (print) | LCCN 2021042728 (ebook)
| ISBN 9780367683085 (hardback) | ISBN 9780367683078 (paperback) | ISBN 9781003136842 (ebook)
Subjects: LCSH: Education--Data processing--Case studies. | Internet literacy--Case studies. | Information resources management--Case studies.
Classification: LCC LB1028.43 .L44 2022 (print) | LCC LB1028.43 (ebook) | DDC 371.33/4--dc23/eng/20211213
LC record available at https://lccn.loc.gov/2021042727
LC ebook record available at https://lccn.loc.gov/2021042728

ISBN: 978-0-367-68308-5 (hbk)
ISBN: 978-0-367-68307-8 (pbk)
ISBN: 978-1-003-13684-2 (ebk)

DOI: 10.4324/9781003136842

Typeset in Bembo
by SPi Technologies India Pvt Ltd (Straive)

Both editors would like to thank the *Australian Research Council Centre of Excellence for the Digital Child* (Grant: CE200100022) for support during the preparation of this volume. The views expressed are those of the authors and editors.

CONTENTS

ILLUSTRATIONS

Figures

Tables

CONTRIBUTORS

Earl Aguilera is an Assistant Professor in the Department of Curriculum and Instruction at California State University, Fresno, in the United States. He is an award-winning teacher educator, as well as an internationally-recognised scholar on issues at the intersection of digital literacies, critical pedagogy and educational equity.

Monica Bulger studies youth and family media literacy practices and advises policy globally. She has consulted on child rights in digital spaces for UNICEF since 2012, and her research encompasses 16 countries in Asia, the Middle East, North Africa, South America, North America and Europe.

Patrick Burton has extensive research and policy experience on child and youth victimisation, school violence prevention and youth resilience. Previously Executive Director of the Centre for Justice and Crime Prevention in South Africa, his research spans 34 countries across Africa, the Middle East, Asia and the Pacific Island Countries and Territories.

Cristobal Cobo is a Senior Education Specialist at the World Bank. He write here in a personal capacity and does not represent the views of the World Bank. He received his PhD from the Autonomous University of Barcelona. Between 2014 and 2019 he was director of the Ceibal Foundation (Uruguay), and spent ten years as a research associate at the Oxford Internet Institute (University of Oxford, UK). His research interests focus on the appropriate and safe use of new technologies in education.

Anne-Sophie Collard is a professor in information and communication sciences at the University of Namur and director of the Communication and Internet Unit at the Research Center for Information, Law and Society (member of the Namur Digital Institute).

Bronwyn Cumbo is a design researcher from Monash University, Melbourne. Her research explores how creative, critical and participatory design approaches can be used to address complex social issues in urban contexts. Her primary interests explore how creative, interactive technologies can cultivate connections between children and other species in natural urban areas.

Emma Day is a human rights lawyer, specialising in children's rights. She has worked for the UNDP, UN Women and UN Global Commission on HIV and the Law in Rwanda, Kenya, Thailand and Indonesia. Since 2017 Emma has led UNICEF East Asia and Pacific Regional Office's work on child online protection.

Mariëtte de Haan is a professor of Intercultural Education at Utrecht University, the Netherlands. Her research focuses on how normative traditions of learning and education relate to more spontaneously created ones, the reimagination of learning in online worlds, and educational responses to datafication in the lives of youth.

Tom De Leyn is a PhD candidate at the Department of Communication Sciences, Ghent University. He is affiliated to the imec research group for Media, Innovation and Communication Technologies (imec-mict-UGent). He is working on a Research Foundation Flanders (FWO)-funded project about youth culture, privacy and media literacy.

Roberto Santiago de Roock is an Assistant Professor of Learning Sciences & Technology at University of California, Santa Cruz. His interdisciplinary work examines the relationships between literacy, technology and abolition under racial capitalism. He primarily does this through ethnographic design work, but also in pioneering critical digital discourse analysis.

Ralf De Wolf is postdoctoral researcher and assistant at an assistant professor at the Department of Communication Sciences and connected to the research group for Media, Innovation and Communication Technologies (imec-mict-UGent), Ghent University, Belgium. His research focuses on the privacy management of children and youngsters, automation, algorithms and inequality.

Rebecca Eynon is a professor at the University of Oxford, where she holds a joint appointment between the Department of Education and the Oxford Internet Institute. Her research examines the relationships between social inequalities, education and technology.

Jérémy Grosman is a PhD student, at the crossroads of History and Philosophy of Science and Science and Technology Studies, currently finishing a thesis on algorithmic practices, titled "Making Algorithms. Finding Optimum. A Philosophical Investigation on Optimization Algorithms".

Jerry Jacques is an Assistant Professor at UCLouvain. His research focuses on practices and competences of users who interact with digital media, with a particular interest in the challenges raised by technical innovations (algorithms, robotics, automation). His research also questions the relations between media, information, and knowledge.

Hyeon-Seon Jeong is a professor of media literacy education at Department of Korean Education and M.Ed. in Digital Media Education, Gyeongin National University of Education, South Korea. Her research focuses on media and digital literacy for children, teachers and parents. She has contributed to scholarly books including Learning Beyond the School.

Niels Kerssens holds a PhD in media studies and is an Assistant Professor at the department of Media and Cultural Studies at Utrecht University, where he leads the special interest group Platformisation of Education. His research investigates the impact of Educational platform Technologies (EdTech) on public values of learning and teaching.

Amie Kim is a visiting researcher at the Center for Media Literacy Research at Gyeongin National University. Her research interests include young people's media culture, media literacy education and children's digital rights. Her recent research looked into young people's experience as digital citizens and primary school students' experiences on YouTube.

Eva Lievens is an Associate Professor of Law & Technology, Ghent University and a member of the Human Rights Centre, the Crime, Criminology & Criminal Policy Consortium, ANSER and PIXLES. She researches the legal impact of technology design and deployment, human and children's rights in the digital environment, and alternative regulatory instruments.

Sonia Livingstone is a professor in the Department of Media and Communications at the London School of Economics and Political Science. Taking a comparative, critical and contextualised approach, she has published 20 books on media audiences, children and young people's risks and opportunities, media literacy and rights in the digital environment.

Marijn Martens is a PhD candidate at the Department of Communication Sciences of Ghent University and connected to the research group for Media, Innovation and Communication Technologies (imec-mict-Ugent). His current FWO-funded PhD project concerns why people (dis)trust Algorithmic Decision Support Systems, adopting a critical and interpretative mixed method research design.

Ingrida Milkaite obtained her PhD from the Law & Technology research group and is a member of the Human Rights Centre and PIXLES at Ghent University, Belgium, as well as ECREA. Her doctoral thesis focused on a children's rights perspective on privacy and data protection in the digital age.

Yeon Ju Oh is a principal manager at the Department of ICT Policy, National Information Society Agency. Specialised in gender studies, science and technology studies and political economy of communication, she critically examines how the society and people's lives are shaped by digitally driven innovation and its discourse.

Juliana E. Raffaghelli is a Ramon y Cajal researcher at the Universitat Oberta de Catalunya (Spain), former collaborator at the universities of Venice, Florence, Trento and the Italian National Research Council. Her research interests relate the technological mediation of professional learning in the global context. Recently, she has been exploring educators' critical data literacies.

Pablo Rivera Vargas is a lecturer at the Department of Didactics and Educational Organization of the University of Barcelona (Spain). Ph.D. from the University of Barcelona. His research activity is related to the analysis of digital inclusion policies and practices in educational contexts inside and outside the school.

Ricarose Roque is an Assistant Professor in Information Science with courtesy appointments in Computer Science and the School of Education at University of Colorado Boulder. She directs the Creative Communities research group and Family Creative Learning project. She designs and studies inclusive learning environments that enable young people to become computational creators.

Katharine Sarikakis is professor of Communication Science at University of Vienna, Austria and heads the Media Governance and Industries Research Lab. She is the Director of the *Jean Monnet Centre of Excellence Communication, Facts and regulation for Europe* (FREuDe). Her research interests focus on transformation, process of decision-making and power relations in media governance.

Neil Selwyn works at Monash University, Melbourne. Twitter: @neil_selwyn.

Mariya Stoilova is a postdoctoral researcher at the Department of Media and Communications, London School of Economics and Political Science. She researches child rights and digital technology, focusing on the risks and opportunities of children's digital media use, data and privacy online, digital skills and pathways to harm and wellbeing.

Ben Williamson is a Chancellor's Fellow at the Centre for Research in Digital Education, University of Edinburgh, UK. He is the author of *Big Data in Education: The Digital Future of Learning, Policy, and Practice* and editor of *Learning, Media and Technology*.

Luo Xiaoming is an Associate Professor working in the Program in Cultural Studies of Shanghai University. Her research interests are urban culture and social space in everyday life, contemporary Chinese science fiction. She is also the chief editor of the online journal *Refeng Xueshu* (2017–2021).

1

LEARNING TO LIVE WELL WITH DATA

Concepts and challenges

Luci Pangrazio and Julian Sefton-Green

Introduction

Digital technologies infuse virtually every aspect of life. From health tracking apps, social media and learning management systems, to the Internet of Things, traffic surveillance, government services, dating, work and education – many of our routines and practices are recorded and analysed somewhere in the world. Scholars and technology experts call this "datafication". Datafication refers to the process in which actions and behaviours are translated into data that can be recorded, sorted or indeed commodified by governments and private companies. The consequences and implications of datafication are immense, extraordinary and unprecedented.

Governments have responded to these challenges in various ways. The European Union has enforced the General Data Protection Regulation (GDPR) to increase data protection and privacy and regulate public competition, while in Singapore the government is addressing "fake news" by taking control of the media. The use of data to manipulate democratic processes has come under scrutiny, most notably around the Cambridge Analytica "scandal" of 2016. While there has been intense concern over the integrity of democratic systems and public information, there has been equal scrutiny of privacy rights especially where they impinge on the security and safety of young people. In many countries around the world there are laws and agencies designed to protect young people, especially from the threats implicit to being online.

Educational responses to the magnitude and range of these challenges have differed around the world. In many respects datafication should be understood as a global phenomenon – not only because of the transnational nature of many of the private companies at work across the countries of the world (i.e., Google), but also because there are extraordinary similarities to the digitisation of everyday life despite different national cultures. In some countries, nationally sanctioned advisory

DOI: 10.4324/9781003136842-1

agencies, such as Australia's eSafety Commission, mix information, advice, and, where possible, regulation and scrutiny of digital activities to protect and uphold rights, safety and education.

However, at the level of school management, curriculum design and/or pedagogy there has not been the same degree of investment or classroom innovation. Some countries have attempted to support the development of online learning systems, while others have introduced social norms, such as the banning of mobile phones. Other individual schools or districts have used digital or other means of communicating with parents and the community to engage interest and support for initiatives that teach young people about the risks and opportunities posed by datafication.

This book collects approaches to teaching and learning about data and datafication from countries around the world. Each chapter explains an aspect of datafication and conceptualises the problem for educational purposes. The authors explore the pedagogic challenges in understanding how students might learn about the problems of data and living in a datafied society and how we might construct curriculum progressions to support deeper engagement. The focus is on methods, approaches, models and theories that conceptualise and implement ways of learning to live well with data. Although all sections of the population are engaged in these forms of learning, the attention in this book is on young people and especially on forms of school-aged education.

The datafication of everyday life

Attending school, joining sporting groups and socialising with friends now typically requires some kind of engagement with digital platforms, and this engagement generates personal data that can be used to profile, track and trace. Personal data refers to any information that can be directly attributed to an individual. Personal data can be drawn from a huge array of sources – from the seemingly insignificant mouse click through to the more expected information of date of birth, address and social security number. It can also take a variety of modes, including numbers, characters, symbols, images, electromagnetic waves, sensor information and sounds (Kitchin, 2014). Indeed, part of the problem in supporting people to understand and engage with digital data is that it is a rather nebulous concept that refers to so many seemingly different pieces of information. As Golumbia (2018, n.p.) explains, personal data is a "much larger and even more invasive class of information than the straightforward items we might like to think". It includes not only what individuals voluntarily share with the platform, but also what can be inferred from this data through probability analytics. Indeed, personal data now drives a multibillion-dollar data broker industry (Couldry and Mejias, 2019; Zuboff, 2019).

There is a growing body of critical scholarship that documents how the collection of personal data not only invades personal privacy, but also, through recommendations, prompts and design trains young people into new routines that increase data generation (Chun, 2016; Langlois et al., 2015). This intensifies a whole host of nascent issues to do with profiling, prediction as well as providing a ready-made way to judge and assess others – for example, social media metrics now shape the way young people think about themselves and others (Gangneux, 2018; Pangrazio, 2019), or in the school domain how teachers and administrations use automated data collection to help gauge the capabilities of their students.

Data and datafication

Datafication is the transformation of digital interactions into a record that can be collected, analysed and commodified (Mayer-Schoenberger and Cukier, 2013). It is made possible by the capacity to capture and translate social phenomena into data and is integral to the business model and functioning of many digital platforms. For students, personal data and metadata is generated through the school-based platforms and social media platforms that they use throughout their day. In some instances, student may be aware of data being generated and captured, like when creating a profile on the school's learning management systems (LMS), but in many others, they are not. For example, while a young person may not even be a Facebook user, if they have visited a site that has a Facebook 'like' button embedded, this third-party cookie can be used to track their online movements. Data have long been used to monitor and regulate populations; however, digital technologies have greatly expanded what can be turned into a data point.

While datafication is a new process, it is also just the next step in the long-standing trajectory towards quantification. Hacking (1990) argues that the drive to quantification has been underway since 'populations' were introduced for statistical analysis in the 17th century. The reliance on numbers was seen as a way of 'taming chance' and providing some form of stability and security in a world that was rapidly changing. This mindset is evident everywhere today. In schools, the use of data may be marketed around optimising performance and managing risk, however, the reality involves new school logics based upon the dataveillance of teachers and students. This creates new routines and regimes for staff and students to follow, which is often characterised by a high degree of manual labour and the emergence of new power dynamics (Selwyn et al., 2021). At the same time, it is easy to see why it might be appealing to the school for the reassurance it can give to teachers and parents.

However, the collection of data about people has implications. Statistics leads to the construct of a 'norm' and mathematical notions of deviance, as well as the ability to predict risk. It both reinforces and intensifies the reliance on data-driven epistemologies (Kitchin, 2014) based upon comparison, pattern recognition, prediction and analytics. Through artificial intelligence and automation, the more 'human' skills of interpretation, reflection and evaluation are displaced. Most digital systems automate feedback (i.e., metrics), as well as curate and personalise content (i.e., recommendations and search). As a consequence, individuals are directed towards particular behaviours and practices, as 'the self is mobilised and activated in response to the calculation to which it is exposed' (Beer, 2016, p. 139). An important purpose of datafication is to enable digital systems to provide an experience that will keep the individual user engaged with the platform longer. Zuboff (2019) argues that datafication is a key tool in enacting 'instrumentarian power' – a new form of power enacted by governments and corporations that can shape and manipulate people in subtle and incremental ways.

Yet big data clearly has benefits for many different sectors in society including education, medicine and science. For example, if used effectively and ethically, data can be used to provide detailed feedback on student performance and provide new and promising insights into the learning process through learning analytics (Gasevic et al., 2015). Data has also become an essential part of modern medicine, providing better health profiles and predictive models that can be used to diagnose and treat disease (Schadt, 2015). As Bhargava et al. (2015) argue, data can also improve civic life. The increasing availability of open data has increased the potential to engage citizens through 'grass roots' innovation. This is dependent, of course, on citizens having the knowledge and skills to access and analyse open data sets. In principle, data can be used for civic empowerment and positive social change. All in all, digital data facilitates many everyday experiences and opportunities, enabling people to build knowledge, swap and share objects, and of course, participate in society.

However, datafication has political implications. Processing data is dependent on the creation of categories and norms, which are often based upon particular social and cultural assumptions. These are assumptions with long, problematic histories. For example, predictive analytics are used in the US to assist with child protection and support, with particular racial groups being unfairly targeted as a consequence (Eubanks, 2017). Without correction embedding digital tools in old systems of power and privilege intensifies social inequalities. For example, some families cannot afford high speed broadband, which means their children's participation in school activities is constrained. Datafication therefore does not affect individuals equally with people of a particular race, religion, income, gender and social status unfairly targeted through data processing.

As data-driven epistemologies are seen as more objective and reliable, datafication changes how we understand social phenomena. It changes how we see ourselves and others shaping the kinds of behaviours and interactions that we engage in. Finally, it has become a new arena for long-standing social justice issues, with implications for decision-making, governance and power in civil society.

...and the responses to it

The challenges brought about by datafication have inspired a range of different responses. From 'top-down' government regulation of tech companies through to 'bottom-up' grass roots activism, all responses have a common focus on changing how data is being 'done' to individuals. More specifically, responses attempt to intercede at different points in the process of datafication – either by decreasing the amount of data that tech companies can collect through legal or technical means or coming up with alternative platforms that are not reliant on the commodification of personal data for their business model. What follows is a brief overview of three main responses to datafication: regulation; technical and tactical; and educational (see Pangrazio and Sefton-Green, 2020). We delineate between educational responses and educational research to explore not only what is being done in schools to support teachers and students develop their understandings of data, but also how the phenomenon is being researched and approached.

Regulation

The notion of rights – particularly digital rights – has been invoked in response to the challenges posed by datafication. In relation to data, the concept has two dimensions: first as individual human rights and second as property rights. In the first instance, a notion of human rights is unsettled by datafication because the rights an individual has in society – for example freedoms of movement, speech and so on – are not legally protected in the digital domain. Traditions from *habeas corpus* to Miranda rights in the US, therefore, do not govern our online interactions.

One response, which has been led by one of the authors in this collection, Sonia Livingstone, is to advocate for the digital rights of the child (Livingstone and Third, 2017). This response makes a strong case for bringing the 1989 UN Conventions on the Rights of the Child (UN CRC) to the digital context, arguing that the internet is typically thought of as a resource for adults and is reflected as such in policy, regulation and ideology. Drawing on key articles in the UN CRC, a special issue in the journal *New Media & Society* (2017), brought together researchers across different fields in the social sciences, to argue that the digital rights for children should cover both participation and provision, as well as protection. More recently, this has developed into the 5Rights Foundation, which seeks to make systemic changes to the digital world to ensure by 'design and default...that children can thrive in a digital world' (5RightsFoundation, 2020).

Regulatory responses seek to prevent unfair and unequal power relations that position the individual in a vulnerable relationship. In this way, they can be an effective reply to data justice issues, as well as encouraging a more interpretive and critical approach to datafied identities and interactions. The language of rights and regulation derives authority from being governed by principles of law that operate in the interests of the individual. For this reason, it is up to the individual to assert their rights, even though these gather greater meaning and significance in regard to the collective. As such, the principle of 'digital rights' needs to be protected by the body politic as much as they need to be asserted by the individual.

Technical and tactical

Where regulatory responses attempt to redress privacy issues and information asymmetries, technical solutions counter these by disrupting the smooth flow of data into the digital economy. In many respects, technical solutions arise in response to a lack of media regulation and are based on the idea that understanding data processing will motivate citizens to try a range of DIY tools and technologies to protect their data. Strategies include software designed to: increase the transparency of data processing; block and obfuscate (Brunton and Nissenbaum, 2015); anonymise and encrypt personal information; and digital tools to help users understand the terms and conditions of digital platforms in order to properly achieve 'informed consent'. Similarly, technology activists are creating more ethical platforms and algorithms to protect personal data from data mining and monetisation. Initiatives such as 'Platform Co-operativism' (Scholz, 2016) and 'Algorithmic Fairness' (Bousquet, 2018) seek to provide alternative designs to the data-driven, profit-based business model that underpins the design of most digital technologies.

Tactical responses are predicated on the notion that the individual understands the politics of their participation and will act to counter or right the situation. This disrupts the idea that users are compliant data 'subjects' within the digital economy, and instead positions them as having the capacity and agency to enact change. However, some tech companies are already a step ahead. Some are circumventing tactical responses by offering tactical resistances either as mode of self-regulation or as goods in the software marketplace. For example, many commercial platforms now offer 'safe' or 'private' browsing modes and/or dashboards that can display to the user what the platform 'knows' about them. The aim is to allay user concerns through increased transparency of data collection. In many ways, this is a rather cynical 'soft' response to data privacy concerns. Nevertheless, tactical responses emphasise protection through individual knowledge and action, rather than regulatory responses that tend to focus on forms of protection being held by society at large.

Educational responses

In general terms, educational responses posit the idea that people should have knowledge about the way that datafication works. However, the actual mechanisms by which that knowledge might be acquired, what it might consist of, and how the possession of it would affect a whole range of behaviours, is often vague. Educational responses thus tend to become distilled into the catchall of 'literacy' (as in 'financial literacy', 'emotional literacy' or 'physical literacy') as terms used to describe the individual's capacity to understand information and the social norms and conventions that surround it, as well as demonstrate this knowledge through writing and/or activity.

In many respects, educational responses are fundamental to both a regulatory and a tactical response as both are – to an extent – dependent on the fact that the individual will have enough awareness and understanding to act – i.e., to claim their individual rights or develop technological responses. Yet the dominant way of learning about datafication today tends to be in the data science tradition, which is dedicated to skill building and information management (Pangrazio and Sefton-Green, 2020). For example, popular programmes such as the Data Literacy Project by Qlik[1], the Data Literacy Foundation's Toolkit[2] and the Data Carpentries workshops[3] are all directed at individuals and/or organisations and focus on building fundamental skills for working and researching with data. However, understanding the more complex shifts brought about by datafication discussed so far call for more than the ability to work with data. It requires critical reflection on how datafication is changing the way we live, learn and work (Pangrazio and Selwyn, 2019). These pose more complex and difficult questions for education as they require not only awareness and understanding of datafication processes, but also the time, space and interpersonal support to reflexively examine personal data practices.

Bearing this in mind, there are decidedly fewer educational programmes working in the critical tradition. Indeed, many of the current critical educational data programmes exist in a realm that is far removed from where education typically takes place (i.e., school). While some of these programmes have been used with community groups (such as the 'Our Data Bodies' project's *Digital Defense Playbook*), incorporating them into school programmes and curricula is another matter entirely. To be useful in a school context, programmes such as these need to connect with what students already know. For example, there is no point in introducing concepts such as 'algorithm' and 'automation' if students have not yet experienced these phenomena first-hand in their everyday lives. Here, the work of Taina Bucher (Bucher, 2018) on Facebook users provides a useful educational approach. Bucher draws on sociocultural neo-Vygotskian principles whereby spontaneous, everyday concepts derived from experience are transformed into 'scientific' ones. Bucher explores the ways that users theorise and make sense of the ways that they are datafied on social media platforms. With this knowledge, educators can more specifically discern what their students know or do not know about datafication processes and how they can be best supported to develop critical understandings and practices.

Research

Recently, a strand of educational research has emerged that takes a critical approach to data policies, governance and implications of digital technologies in schools. This approach includes research that examines digital technologies from *within* the school setting through ethnographic accounts of how students, teachers and administrators engage with digital infrastructures (Selwyn et al., 2018; Selwyn et al., 2021); and from *outside* the school, through critical analysis of the data policies and governance of these technologies (Manolev et al., 2019; Perrotta et al., 2020; van Dijck et al., 2018). Using a similar critical approach, recent work by Kerssens and van Dijck (2021) has expanded the analytical frame to consider the tensions that arise when commercially owned digital platforms are inserted into national public-school systems. A range of methodologies are used to do this multi-disciplinary work from digital ethnography, infrastructural mapping, digital 'walkthrough' methods (Light et al., 2018) and critical policy analysis.

However, educational research into the effects of datafication has tended to take a distanced perspective on the phenomena, researching policies or terms and conditions. This is perhaps inevitable given how the recent and ongoing pandemic drew attention to the role of educational technologies as societies desperately tried to grapple with enforced changes to the institution of schools. A recent comprehensive overview of the 'EdTech' ecosystem (Williamson and Hogan 2020) offers a comprehensive analysis of the relationship between big business, national and international education policy and the complex ways that forms of surveillance and datafication provide a commercial rationale for the penetration of Silicon Valley based industries into the hitherto more regulated and accountable school systems. This kind of analysis is needed and contributes to public debate; however, as a mode of enquiry, it benefits from a top-down "platform gaze". Following David Beer's (2018) idea of the "data gaze", and his analysis of social imaginaries we are suggesting that the analytical phenomenon of the platform has rather overwhelmed and preoccupied research in this area, perhaps at the exclusion of alternative and complimentary approaches.

The critical perspective tends to look downwards as it were from the point of power analysing the ripple effect of high-level decisions made by national governments and supranational corporations. We absolutely support and espouse this perspective but the debate in general needs to be able to account for the effects of take-up, use and meaning making from a much wider range of social actors to show how policy decisions are actually implemented in practice. This will reveal the impact of modes of surveillance and forms of datafication on young people, their families, teachers and schools and schooling in general rather than only looking at the effects from a single privileged perspective. A recent collection entitled 'COVID-19 from the margins' (Milan, Trere, and Masiero, 2021) exemplifies this more comprehensive approach showing how forms of datafication are negotiated and resisted in a range of everyday settings by diverse populations in local contexts. The platform gaze is a necessary and important contribution to scholarship and public debate, but one of the aims of this book is that platformisation should be contextualised through study of vernacular practices, tactical interventions and local interpretations. One of our ambitions with this collection is that it should stimulate further research into the actual *practices* of datafication by diverse communities

Why this book now?

While there are some encouraging signs, recent educational research both *into* and *about* datafication suggest there is clearly a need for a more sustained and focused response from the field of education. In particular, educational research *into* datafication has been characterised by a tendency to adopt the "platform gaze", where the focus is on legal and technical discourses around "Terms of Agreement" and policy responses. While this work is clearly important, it is also essential that we understand how educational stakeholders respond to the realities of datafication – that is living, teaching and learning in datafied systems. Datafication introduces new regimes, routines and hierarchies into schools that not only change the values and beliefs of teachers and staff, but also the focus and priorities of curriculum and assessment.

Of course, schools and educational institutions now rely on digital platforms to conduct a whole host of daily activities. From delivering content to students, to processing attendance data and managing staffing requirements – there is an educational platform designed to address all schooling needs. However, educational platforms, like other digital platforms, are powered by data. The more teachers and schools depend on platforms to do their work, the more data is generated, collected and used by technology companies. Some of this data is personal and sensitive and concerns student health and wellbeing, yet this can be mined under the guise that it will customise learning and streamline schooling processes. Commercial access to student data by educational platforms is therefore discursively constructed as integral to efficiency, knowledge production, and even good citizenship (Yu and Couldry, 2020).

Many questions emerge from this situation. Teachers and administration often presume that all students can access and download software equally – they are after all assumed to be 'digital natives'. However, the reality is often quite different. Not only are young people *not* equally adept at using technology or inherently better at it than adults (Livingstone, 2009; Selwyn, 2009), but it appears schools and communities are often not prepared for the complex challenges that emerge from the generation of digital data. For example, even in Australia, which is a relatively wealthy country, many students still do not have access to adequate digital technologies or the resources and support to learn how to use them properly. One key theme then is the kind of knowledge and understanding schools need about their own practices of datafication and the impact of platformisation on how they traditionally have worked.

The second key theme is the effects of datafication on ideals of civic and social identity for students themselves as they inherit the challenges of living in a platform society. Given the complexity of datafication processes, it is perhaps not surprising that educational responses have struggled to keep pace. Indeed, there are number of complexities that educational responses are still grappling with. First, due to the opaque, black-boxed nature of datafied systems (Pasquale, 2015), it is difficult to materialise a text for deconstruction and critical analysis – as the tradition of criticality demands (Pangrazio and Sefton-Green, 2020). Second, most of the current educational programmes are based on the idea that awareness and understanding will compel the individual to act. This is not always the case. Unlike other cultural and political systems that are the topic of critical analysis, datafication is not only deeply intertwined with our socialities, but its implications also mean greater efficiency and precision when it comes to insights on the self and others, much of which is experienced 'after the fact'. Third, if one *were* motivated to act then what would acting or resisting actually mean? Given the network effects of mainstream digital platforms (Belleflamme and Peitz, 2016), typically resistance tends to mean opting out or compromising use, rather than finding viable alternatives. Indeed, there are so few alternatives that it is difficult for a critical education *not* to invoke paranoia given it means compromising social experiences in some way.

Clearly, teachers, policymakers and educational bureaucrats need to see the importance of critical data education and provide adequate time and resources to support its design and integration with current curriculum if it is to be effective. We need to know how different educators conceptualise the social implications of datafication and what is at stake for learners and citizens as educational institutions try to delineate what datafication will mean for the next generation. The chapters in this book offer powerful and practical approaches to critically navigating one of the most complex issues of our time. This book aims to provide teachers, students and policy makers with the information they need to know about datafication and the kinds of educational interventions that work.

Overview of chapters

This collection is focused on educational research and responses to datafication. Some of the case studies take place in schools, while others look beyond the school to consider issues of regulation, policy analysis and cultural politics. Some chapters are focused specifically on datafication in education, but others look at datafication broadly yet in an educational setting. Despite these variations, the common themes across the collection are datafication and education, with each chapter exploring and finding innovative strategies to counter the various ways these two concepts intersect. The collection explores five sub-themes: Policy; Schools and Teachers; Algorithms; Platformisation; and Regulation. Each of these areas is developed by two chapters from different countries from around the world.

In Chapter 2, Rebecca Eynon explores the implications of using digital trace data for educational purposes. She examines how datafication is influencing the three main interrelated roles of schools through qualification, socialisation and subjectification, as argued by the philosopher, Gert Biesta. Eynon considers the implications that digital trace data is likely to have for social change and the transformation or reproduction of existing inequalities. The chapter then draws attention to the kind of education system a society wants and needs in an era characterised by use of digital trace data. Given the problems with the current status quo and the likely future for schools, the final part of the chapter focuses on educational interventions that could counter these trends. In particular, the chapter stresses the importance of data literacies and the need for participatory design to ensure educators and students begin to create alternative data futures that lead to more positive outcomes for all. The chapter concludes by arguing that these strategies will help schools to directly contribute to the current needs of society and also to provide places of reflection and action that may challenge and change the status quo.

In Chapter 3, Cristobal Cobo and Pablo Rivera-Vargas explore how different education systems in South America managed the challenges associated with videoconferencing services during the COVID-19 pandemic. The outbreak triggered an explosive use of digital technologies to enable remote learning, making Zoom Video Communications, Inc. (the global provider of video conferencing) one of the most popular educational technologies. Although Zoom has enabled remote learning in many parts of the world, it has faced several global concerns including sharing private information with third parties, the adoption of dubious end-to-end encryption processes, and exposing users to unauthorised intimidation calls. Cobo and Rivera-Vargas carried out a virtual ethnography and discourse analysis to investigate the extent to which governments alerted their citizens to the potential risks to their privacy. They analyse 48 websites of public government institutions from eight countries – Argentina, Brazil, Colombia, Chile, Ecuador, Paraguay, Peru and Uruguay. Their findings emphasise the need to review and adjust current regulations. In particular, they call for the strengthening of existing institutional and governmental capacities to effectively execute digital governance. This chapter considers the role of regulatory responses and will be of particular interest to educational policy makers interested in understanding not only the opportunities of digital technologies, but also the unintended consequences.

Chapters 4 and 5 in the collection shift the focus to key stakeholders in the datafication of education, namely schools and teachers. Following the emerging tradition of 'critical data studies', Neil Selwyn, Luci Pangrazio and Bronwyn Cumbo report on their qualitative investigation into the digital data practices of three secondary schools in the Australian state of Victoria. In particular, the chapter focuses on the important question of who gets to 'do' data within school contexts (and, conversely, who does not) by drawing on in-depth interview data with teachers, IT specialists and administrators in each school. Belying outward appearances of being successful 'data schools', five distinct hierarchies of data-using staff are identified – each aligned with a number of notable reconfigurations of power and redistributions of agency. The chapter discusses the ways in which pressures to use data are entwined with wider reformations of teacher identity and teacher professionalism. In particular, the authors consider how any conferred benefits of 'doing data' seem delineated by a range of significant factors – not least teacher gender, status, disciplinary background and career stage. They conclude the chapter by arguing that the dominant discourses of the educational benefits of data-driven schooling need to be challenged – if not reconsidered altogether.

Juliana Raffaghelli takes up the challenge of supporting educators' data literacy in the fifth chapter. Her focus is on supporting educators' understandings of digital data so they, in turn, can address issues of social justice in their various educational institutions. Raffaghelli begins the chapter by acknowledging that there is not a universal perspective on educators' data literacy that institutions and the system can embrace. Instead, she argues that professional knowledge and practices and the development of these need to embrace an approach to data literacy that is evolving, situated and explorative. Working with almost 300 participants across 12 workshops in both English and Spanish speaking communities, Raffaghelli explores how educators' data literacy develops over time, as well as how they are related to their faculties, students and other stakeholders' efforts to make sense of the data politics, epistemologies and practices. The workshops involved the use of 'hermeneutic circles', where the educators' conversations triggered understandings around their literacies to explore situated data epistemologies and practices. She concludes the chapter by reflecting on the educators' engagement with situated data practices and how these can be used to promote relevant professional development.

Chapters 6 and 7 examine the platformisation of education, as well as its inherent relationship to datafication. Earl Aguilera and Roberto de Roock begin Chapter 6 by arguing that datafication has a tendency to overlook the interactional and dynamic nature of the datafication process, as well as the complex responses of individuals. The chapter unpacks some of these complexities in the context of online LMS through the lens of *proceduralisation*. This term is used to describe the ways that ideological constructions (such as social practices, cultural viewpoints, and implicit biases) are embedded in and circulated through the designs of technologies. Using mediated discourse analysis, they investigate the Canvas LMS, highlighting how the procedural design and applied use of the platform evidences a particular ideological model of teaching and learning. They base their analysis on a series of qualitative interviews with students in the western US focusing on their beliefs, experiences and strategies for navigating the LMS, as well as their own analytic walkthroughs of the software itself. The chapter concludes by discussing the pedagogical implications for the material and ideological responses of those involved with datafication through learning systems.

In Chapter 7, Niels Kerssens and Mariëtte de Haan explore the instrumental rationale that underpins the platformisation of education. They argue that critical interventions and policies would benefit from a thorough examination of digital platforms in education, in particular how these have been received and contested. They examine how educational platformisation in the Netherlands is the result of at least two decades of educational reforms grafted on *instrumental rationality* and map the counter-discourse that has emerged from public stakeholders and interventions in the educational field that rally against an instrumental rationale for digitalisation of educational practice. This budding counter-discourse, they argue, presents a tipping point which can be further sustained through interaction with earlier fundamental critiques by educational scientists who also counter the instrumentalist and technocratic uptake of technology. The chapter concludes by reflecting on what the analyses means for the conceptualisation of the platformisation of education, and what such a perspective might mean in terms of advice and guidelines for policy makers and practitioners.

Chapters 8 and 9 can be read as a pair as both sets of authors used a common stimulus to develop and reflect upon a teaching unit to develop and enhance understandings of datafication. In Chapter 8, Jérémy Grosman, Jerry Jacques and Anne-Sophie Collard all based at the University of Namur in Belgium, describe and evaluate a unit of work they developed to teach young people about YouTube recommendation algorithms. In the following chapter, Hyeon-Seon Jeong, Yoeonju Oh and Amie Kim, representing a mixture of university research, NGO and government perspectives, further develop the principles of this simulation. In both cases, students are required to take up the positions of the YouTube channel engineers and even perform the role of the algorithm itself in order to develop and support both their understanding of the determining power of algorithms in affecting consumer choice, as well as opportunities for critical intervention and resistance that might be developed by the young people as a consequence of such educational experiences.

Both chapters offer an evaluation of the teaching unit trying to expose the challenges in how to get students engaged in understanding what it might mean for them to be datafied. Both chapters are interested in the ways in which personal experience can be used to give the student a stake in this kind of learning, but also the political and social consequences of this kind of media education so that abstract learning about algorithms can be translated into real life personal responses. Both chapters describe the basis on which the simulation was devised, and the challenges involved in implementing this kind of game as a way of teaching about such important and sensitive social concerns. The chapters are an interesting contrast not just because they deal with different age groups and come from opposite sides of the globe, but because what counts as meaningful responses and learning for the educators and the students suggests a new kind of more universalised curriculum focus. Both sets of authors are open about the fact that this kind of intervention is novel for both students and educators and suggest a number of challenges for such programmes to be scaled and mainstreamed at the same time as underscoring the political, social and ethical need for these kinds of projects.

Many of the authors involved in this book share social and political norms in respect of privacy and the threat that datafication poses to our sense of self. The penultimate chapter by Luo Xiaoming from the University of Shanghai does not necessarily start from that premise as the relationship between individual rights, subjectivities and the state are reconfigured differently in China. The chapter takes as its premise a social context where datafication is already rampant and where the purposes of educating individuals to understand and respond to knowledge about them by both state and commercial actors could not suggest forms of resistance and protection that animate many other chapters in the volume. The study describes how university students engage with forms of cynicism, resignation and acceptance and where teaching about the nature of surveillance and control has different kinds of objectives in terms of the maintenance of social trust. The chapter is interested in configuring the relationship between a form of social science where understanding and knowledge can be harnessed for progressive positive ends and the critical purposes of social science for citizens to challenge and question the regimes in which they find themselves. This does not suggest a form of liberatory education as perhaps earlier chapters in the volume might suggest and acts as an interesting corrective to what might be possible in teaching and learning about data in the future for many countries around the world.

The final chapter, 11, written by a large international collaboration led by Sonia Livingstone however does suggest a more universal experience of privacy as articulated by children from around the world. Throughout the whole volume there is a recognition that understandings of datafication are a challenge for policy regulation as much as for education. Obviously the more educated citizens are about data, the more they may demand of their regulators. However, more transparent data regimes in different countries will affect the need for different kinds of educational interventions. This chapter stems from a child rights approach, comparing and contrasting children's understanding of digital literacy, digital rights and privacy across a diverse range of social and political contexts. It describes some of the current limitations that govern both the definition and implementation of digital rights for children and ends the book arguing for the need for a qualitatively different digital rights regime. The authors argue that in order to protect children's interests and to ensure that the potentially negative and intrusive effects of datafication can be properly understood and controlled, there is a need for forms of governance that acknowledge and underwrite children as key constituents in contemporary society.

Notes

1 See: https://thedataliteracyproject.org
2 See: https://dataliteracyfoundation.org
3 See: https://datacarpentry.org

References

5RightsFoundation. (2020). Introduction Retrieved from https://5rightsfoundation.com

Beer, D. (2016). *Metric Power*. London: Palgrave Macmillan.

Beer, D. (2018). *The Data Gaze: Capitalism, Power and Perception*. London: SAGE Publications.

Belleflamme, P., & Peitz, M. (2016). Platforms and network effects. In L. Corchon, M. Marini, & E. Elgar (Eds.), *Handbook of Game Theory and Industrial Organization*: Elsevier.

Bhargava, R., Deahl, E., Letouze, E., Noonan, A., Sangokoya, D., & Shoup, N. (2015). Beyond data literacy: Reinventing community enagement and empowerment in the age of data. *Data Pop Alliance White Paper Series*. Retrieved from https://datapopalliance.org/item/beyond-data-literacy-reinventing-community-engagement-and-empowerment-in-the-age-of-data/

Bousquet, C. (2018). Algorithmic Fairness: Tackling Bias in City Algorithms. *Data Smart City Solutions*. Retrieved from https://datasmart.ash.harvard.edu/news/article/algorithmic-fairness-tackling-bias-city-algorithms

Brunton, F., & Nissenbaum, H. (2015). *Obfuscation: A User's Guide for Privacy and Protest*. Massachusetts, CA: The MIT Press.

Bucher, T. (2018). *If...Then: Algorithmic Power and Politics*. Oxford, UK: Oxford University Press.

Chun, W. (2016). *Updating to Remain the Same: Habitual New Media*. Cambridge, MA: The MIT Press.

Couldry, N., & Mejias, U. A. (2019). *The Costs of Connection: How Data is Colonizing Human Life and Appropriating it for Capital*. Stanford, CA: Stanford University Press.

Eubanks, V. (2017). *Automating Inequality: How High-Tech Tools Profile, Police, and Punish the Poor*. New York: St Martin's Press.

Gangneux, J. (2018). Logged in or locked in? Young adults' negotiations of social media platforms and their features. *Journal of Youth Studies*. doi: 10.1080/13676261.2018.1562539

Gasevic, D., Dawson, S., & Siemens, G. (2015). Let's not forget learning analytics are about learning. *Tech Trends*, *59*(1), 64–71.

Golumbia, D. (2018). We don't know what 'personal data' means. *Uncomputing*. Retrieved from https://www.uncomputing.org/?p=1983

Hacking, I. (1990). *The Taming of Chance*. Cambridge, UK: Cambridge University Press.

Kerssens, N., & van Dijck, J. (2021). The platformization of primary education in The Netherlands. *Learning, Media and Technology*. doi: 10.1080/17439884.2021.1876725

Kitchin, R. (2014). *The Data Revolution: Big Data, Data Infrastructures, & Their Consequences*. London: Sage.

Langlois, G., Redden, J., & Elmer, G. (2015). Introduction: Compromised data—From social media to big data. In G. Langlois, J. Redden, & G. Elmer (Eds.), *Compromised Data—From Social Media to Big Data*. New York and London: Bloomsbury.

Light, B., Burgess, J., & Duguay, S. (2018). The walkthrough method: An approach to the study of apps. *New Media & Society*, *20*(3), 881–900.

Livingstone, S. (2009). 'Digital Natives': A Myth? In R. Das & C. Beckett (Eds.), *A POLIS paper*. Retrieved from http://www.lse.ac.uk/media@lse/Polis/Files/digitalnatives.pdf

Livingstone, S., & Third, A. (2017). Children and young people's rights in the digital age: An emerging agenda. *New Media & Society*, *19*(5), 657–670.

Manolev, J., Sullivan, A., & Slee, R. (2019). The datafication of discipline: ClassDojo, surveillance and a performative classroom culture. *Learning, Media and Technology*, *44*(1), 36–51.

Mayer-Schoenberger, V., & Cukier, K. (2013). *Big Data: A Revolution That Will Transform How We Live, Work and Think*. London: John Murray.

Milan, S., Trere, E., & Masiero, S. (2021). *Covid-19 From the Margins: Pandemic Invisibilities, Policies and Resistance in the Datafied Society*. Amsterdam, The Netherlands: Institute of Network Cultures.

Pangrazio, L., & Selwyn, N. (2019). 'Personal Data Literacies': a critical literacies approach to enhancing understandings of personal digital data. *New Media & Society*, *21*(2), 419–437. doi:10.1177/1461444818799523

Pangrazio, L. (2019). Technologically situated: The tacit rules of platform participation. *Journal of Youth Studies*, 1–19. doi: 10.1080/13676261.2019.1575345

Pangrazio, L., & Sefton-Green, J. (2020). The social utility of 'data literacy'. *Learning, Media and Technology*, *45*(2), 208–220. doi: 10.1080/17439884.2020.1707223

Pasquale, F. (2015). *The Black Box Society: The Secret Algorithms that Control Money and Information*. Cambridge, Massachusetts: Harvard University Press.

Perrotta, C., Gulson, K. N., Williamson, B., & Witzenberger, K. (2020). Automation, APIs and the distributed labour of platform pedagogies in Google Classroom. *Critical Studies in Education*. doi: 10.1080/17508487.2020.1855597

Schadt, E. (2015). The role of big data in medicine. *McKinsey Quarterly*. Retrieved from https://www.mckinsey.com/industries/pharmaceuticals-and-medical-products/our-insights/the-role-of-big-data-in-medicine

Scholz, T. (2016). *Platform Cooperativism: Challenging the Corporate Sharing Economy*. New York: Rosa Luxemburg Stiftung. Retrieved January 26 2021 from https://bit.ly/2bSWItd

Selwyn, N. (2009). The digital native – myth and reality. *Aslib Proceedings*, *61*(4), 364–379.

Selwyn, N., Nemorin, S., Bulfin, S., & Johnson, N. F. (2018). *Everyday Schooling in the Digital Age*. London and New York: Routledge.

Selwyn, N., Pangrazio, L., & Cumbo, B. (2021). Attending to data: Exploring the use of attendance data within the datafied school. *Research in Education*. doi: 10.1177/0034523720984200 van Dijck, J., Poell, T., & de Waal, M. (2018). Education. In J. van Dijck, T. Poell, & M. de Waal (Eds.), *The Platform Society*. Oxford, UK: Oxford Scholarship Online.

Williamson, B., & Hogan, A. (2020). *Commercialisation and privatisation in/of education in the context of Covid 19*. Education International. Accessed 21 October 2021, Retrieved from https://issuu.com/educationinternational/docs/2020_eiresearch_gr_commercialisation_privatisation?fr=sZDJkYjE1ODA2MTQ

Yu, J., & Couldry, N. (2020). Education as a domain of natural data extraction: Analysing corporate discourse about educational tracking. *Information, Communication & Society*. doi: 10.1080/1369118X.2020.1764604

Zuboff, S. (2019). *The Age of Surveillance Capitalism: The Fight for a Human Future at the New Frontier of Power*. New York: Public Affairs.

2

DATAFICATION AND THE ROLE OF SCHOOLING

Challenging the status quo

Rebecca Eynon

The datafication of education

There is a long history of data collection in education, and investment, interest and excitement about the potential of data grows year on year (Watters, 2014). At present, it is helpful to consider two broad kinds of digital data collected about young people and their education. The first consists of information about student demographics, course enrolment, and an array of summative and formative assessment data that schools are increasingly required to collect to meet the accountability demands placed upon them by governments. The vast majority of this data is collected via examinations, audit and accounting procedures, responses to surveys and audits, and large-scale government data gathering activities. Examples of such data include the OECD's Programme for International Student Assessment, and well-established national programmes such as the National Pupil Database (NPD) in the UK.

Over the past decade or so, there has been increasing attention on a second, new kind of data, often described as big data or, more accurately, digital trace data. In essence, this is the petabytes of very fine-grained, real-time data generated when an individual uses digital technology for learning. Systems that produce such data include learning management systems, digital games, augmented and virtual reality systems, intelligent tutors, adaptive assessments, social networking platforms, wikis, search engines, and blogs; and are encountered via a growing range of networked wearable, personal and mobile devices (Fischer et al., 2020; Luckin et al., 2016; Romero and Ventura, 2020). These interactions can take place in what are sometimes described as blended learning contexts (i.e., as a part of teaching that takes place face-to-face), or can be a primarily digital experience in or out of school.

The increasing use of digital trace data has significant implications for understanding the purposes of education. This chapter begins by providing an overview

DOI: 10.4324/9781003136842-2

of some of the key features of the datafication of education. It then examines how the use of digital trace data is influencing the central roles of schools, drawing on the work of Gert Biesta; and highlights how the current use of such data is likely to compound existing inequalities. The chapter then provides a brief account of current responses to digital trace data both within and outside education and argues that these are inadequate. The final part of the chapter argues for the need to reconfigure data use in schools via digital literacy and participatory design to help to challenge the status quo.

Characteristics of digital trace data

Both kinds of data are in some sense big. Datasets of each type often contain millions of data points, and each have implications for educational policy and practice. However, some characteristics of digital trace data are distinct from more traditional forms of educational data. The most well cited of these are the 3Vs – "Huge in *volume*, consisting of terabytes or petabytes of data; High in *velocity*, being created in or near real time; Diverse in *variety* in type, being structured and unstructured in nature, and often temporally and spatially referenced" (Kitchin, 2014: 67, drawing on Laney, 2001).

Due to these characteristics, digital trace data offers a significant array of real-time, fine-grained data in ways that are not typically available with other methods. In relation to learning, for example, interactions between teachers, students and their peers, patterns of engagement with course material, learners' emotional and physical states can all be captured at a level of detail that may be difficult to observe in the classroom (Fischer et al., 2020; Luckin et al., 2016; Romero and Ventura, 2020). These insights are then used in schools to detect, understand and intervene in learning in some way. In some cases, such data might be used as part of an automated teaching and learning system (e.g., intelligent tutoring systems or adaptive assessments) or the basis of systems to inform students or teachers about progress (e.g., via LMS dashboards) that can aid education decision-making. These are sometimes retrospective (e.g., mapping activities over the term) but are increasingly part of predictive systems (e.g., likely exam performance, and drop-out risk) with some systems also offering specific interventions to address these issues (Fischer et al., 2020; Hakimi et al., 2021).

Though learning is a core focus, there is a growing number of broader uses of digital trace data relevant to research and policy such as informing resource allocation, evaluating interventions, timetabling and accountability procedures. Indeed, such data can be used by individuals, schools, educational policy-makers and the commercial sector in myriad ways, and can sit alongside and sometimes be combined with existing data-driven processes such as school inspections, school league tables and policy comparisons (Ozga, 2016; Williamson, 2021).

Precisely what this kind of data does and does not offer is the subject of much academic debate. Yet in policy and commercial discourse, digital trace data is typically framed as having almost magical properties, despite the realities being far more complex. Debates around big data are having a significant impact on understandings of what matters in social life and the knowledge claims society values. As Markham notes, big data

> functions as a powerful frame for discourse about knowledge – both where it comes from and how it is derived; privileges certain ways of knowing over others; and through its ambiguity, can foster a self-perpetuating sensibility that it is incontrovertible, something to question the meaning of, or the veracity of, but not the existence of.
>
> (Markham, 2013, n.p.)

This is clearly apparent in education systems where digital trace data is positioned as the solution for all current challenges in education. It is, in this sense, the latest in a long line of 'technical fixes' for education where fixing education is primarily about making the delivery of education more efficient and effective (Robins and Webster, 1989). As Ozga notes, for decades "there is a normalisation of digital data work within education / schooling, where it is conventionally understood and defended as the basis of improvement" (Ozga, 2016: 70). Digital trace data has added to this trajectory. Within this framing big data is positioned as having some kind of transformative power despite there being very little clarity about what precisely in education needs to be transformed or why such data is the answer to such problems.

This way of framing digital trace data is largely due to the efforts of the commercial sector. The role of the commercial sector in education, and the promotion of different kinds of data to support effectiveness and efficiency measures is not new. However, it is perhaps intensifying with the move to focus on digital trace data and related debates around Artificial Intelligence (AI) (Davies et al., 2020) where digital trace data is being used to make education function more like markets; i.e., in making education more accountable, measurable and comparable. Importantly, the use of digital trace data was first used and promoted by the commercial sector to enhance profit and make efficiency gains; and these neoliberal logics can be seen in many spheres of public life including education (Crawford et al., 2014).

The role of the commercial sector

As with any apparently new intervention, the ways that this new kind of data is being taken up, thought about and used in education has been significantly shaped by a complex network of actors that includes multinational businesses, the world bank, OECD; regional states and organisations like the EU and global philanthropies alongside national and local actors (Ball, 2017). Significant parts of these networks are made up of commercial organisations who are influencing policy-making in ways that are likely to be beneficial to them.

These actors, then, offer both discursive and practical ways to influence policies in nation states, while also encouraging a more global agreement on the need for reform in education institutions (Ball, 2017). For example, Pearson, who openly seek to be "an active participant in national educational policy conversations"

(Ball, 2017: 44), produces a series of software and applications available to schools; and also provide analytics insights to governments (Williamson, 2021). Companies such as Pearson not only offer the specific systems for schools, but also "own the algorithms and analytics required to make sense of those data" (Williamson, 2017: 105).

The growing prominence and power of the commercial sector has significant implications for education, as certain kinds of values are promoted, typically that support commercial agendas but also that assume a strong economic purpose for education (Ball, 2017). Furthermore, through the increasing power of these actors, questions of education are no longer focused on the purposes or role of schools. Instead, they have been replaced with questions simply about the best process of delivering education through the use of advanced data analytics (Biesta, 2015). However, such data is of course, not neutral and its use has implications for what schools 'do' and what their relationship is to society.

The purposes of schooling in an era of datafication

A strong and growing body of work has highlighted a number of problematic implications of the datafication of education. These include issues of privacy, surveillance, performativity, instrumentalism and governance (see Ball, 2017; Porter, 1994; Jarke and Breiter, 2019; Williamson, 2021). Digital trace data intensifies these and other concerns in part due to its size, velocity, temporality and relational nature (Kitchin, 2014). The implications of digital trace data are not just a quantitative change (in terms of amount and nature of data collected about education) it also has potentially different qualitative implications (Webster, 2014). As Porter notes, new forms of data "must ask not only about their validity but also about how the world might be changed by adopting new forms of quantification" (Porter, 1994: 404).

Empirical studies of how these data-intense systems are used and taken up by educational institutions, are, as illustrated in other chapters in this book, an essential component of better understanding the social and educational implications of digital trace data. This chapter aims to complement these empirical studies within the context of exploring questions about the purposes of education (see also Eynon, forthcoming).

There are, of course, multiple ways of conceptualising the purpose of education. Here the discussion uses Biesta's very helpful framework (Biesta, 2015). He suggests a focus on three interrelated purposes or roles of education: qualification, socialisation and subjectification. Qualification functions provide young people,

> with the knowledge, skills and understandings and often with the dispositions and forms of judgement that allow them to 'do something'. [...] [This] can range from the very specific (such as in the case of training for a particular job or profession, or the training of particular skill or technique) to the much more general (such as an introduction to modern culture, or the teaching of life skills etc).
>
> (Biesta, 2015: 19–20)

Socialisation, in other words, the ways that "we become part of particular social, cultural, and political 'orders'" (ibid, p.20) is another important function of schooling. This can be both intentional and unintentional and can be both positive and negative. In this way, schools are important in the "continuation of culture and tradition" (Biesta, 2015: 20).

Subjectification is "the process of becoming a subject" (ibid, p.21) where the focus is on "the kinds of subjectivity that are made possible as a result of particular educational arrangements and configurations [...] that allow those educated to become more autonomous and independent in their thinking and acting" (Biesta, 2015: 21).

Biesta argues that reflecting on the optimum combination of the three roles of education provides a useful way to promote conversations about the kinds of education a society wants at different times and contexts and moves society away from a focus on data and measurement (Biesta, 2015). How then could the use of digital trace data influence these roles of a school?

Data and qualification

All three purposes of schooling (qualification, socialisation and subjectification) are clearly important roles of education, yet in many countries there is a tendency to focus increasingly on the economic aspects of qualification (Biesta, 2015). Positioning schools in a way that prioritises societal needs, particularly those of the economy, has narrowed perspectives about what schools are for, from providing a democratic education towards an economic agenda of developing skills for economic gain in the job market (Powers et al., 2016). Educational outcomes (and indeed the curriculum as a whole) have become relatively narrow as the focus is on the economic needs of society (Biesta, 2015).

This trend of narrowing outcomes is exacerbated by the increasing use of digital trace data due to the ways it can:

1) narrow the curriculum through measurement;
2) change the content of the curriculum; and
3) aid in standardising the curriculum.

Measurement is a process, where we "cut up the world" into discrete entities. For any computational analysis to take place, "the continuous flow of our everyday reality [is transformed] into a grid of numbers that can be stored as a representation of reality" (Berry, 2011: 1–2). Data does not straightforwardly represent reality (Borgman, 2015; Espeland and Stevens, 2008). Data are socially constructed representations that are shaped by the expertise and background of the person or people engaged in data collection and analysis, their current motivations for using the data, shaped by temporal and geographical circumstances (Boellstorff, 2013; Borgman, 2015).

What is assessed in education has long been a contested topic, both in terms of the validity of what is measured, and what counts as educationally meaningful

(Ball, 2017). With the increasing use of novel forms of trace data in education, data scientists are making these choices to determine which measures to include to best fit to a model and what the most appropriate outcome measures should be. Indeed, sometimes such choices are made by the algorithm as opposed to a human. In both cases, it is not always clear what measures in data models really represent, even if they improve model fit. Similarly, the outcomes of such models are typically very narrow to capture a complex aspect of schooling, but such assessments were only ever intended to be a proxy for wider achievements (see Luke, 2009). Nevertheless, the measures then become encoded as the central goals of education and prioritise aspects of education that can be measured easily rather than aspects of education that society values but are far more difficult to measure (Biesta, 2015).

The designers of data-intense systems for education have an increasing opportunity to shape the curriculum in ways they see as appropriate because they are building the system. As the majority of such systems are built by the commercial sector their content is based on topics that the data and EdTech companies think matter. As we have seen throughout educational history, it is not just *what* knowledge is most valued, but *whose* knowledge is most valued when thinking through questions of power in education (Apple, 2012). Data scientists and the commercial sector have more power than ever in determining what kinds of knowledge, whose histories and whose content counts. This is not only about narrowing the curriculum, but also standardising it. Monolithic approaches or scaled interventions are something education technology companies desire as this way they can maximise profit and reduce costs. Such activities, automate and standardise knowledge, curriculum and pedagogy in ways that are highly problematic (Saltman, 2016).

The increasing focus on a narrow set of scores has multiple impacts on education, often as they are used not only to test students, but also as part of accountability mechanisms for teachers and schools. This leads to risks of teaching to the test (Selwyn, 2019), impacting pedagogical choices of teachers both in terms of how they teach and also what they teach (Roberts-Holmes and Bradbury, 2016) and encourages performativity and learning for the test in students (Roberts-Holmes and Bradbury, 2016). These challenges are not new, but data-driven systems are reinforcing and intensifying the long-entrenched emphasis on education as an exam-oriented activity and closing down any possibility of moving away from that negative kind of approach to education. Importantly, these trends and concerns with particular parts of the curriculum or particular kinds of knowledge have wider implications than what students learn. As Biesta notes, "students do not just learn from the content we provide them with but also from the ways in which we provide them with this content" (Biesta, 2019: 4) and this has implications for socialisation and subjectivity.

Data and socialisation

Socialisation draws attention to the ways that young people are socialised into existing orders, the ways that schools "(re)present cultures, traditions, and practices, either explicitly but often also implicitly" (Biesta, 2015:20).

Data can shape how individuals think about these cultures, traditions and practices because it encodes and shapes how actors think about themselves and their relationships to others. Well-established and important examples include the case of gender or intelligence, where data is used to define neat categories and boundaries from messy and complex realities (Espeland and Stevens, 2008). In doing so it shapes people's expectations about the world and their identity within it, it assists in a process 'making up people' (Espeland and Stevens, 2008, drawing on Hacking, 1999). This process influences everyone, but is particularly pertinent for young people who are determining who they are in the world (Davies and Eynon, 2013).

When teaching and learning is to some degree automated, as is typically the case with the use of digital trace data, it is likely to have effects on how students think about themselves and the expectations society has for them (Apple, 2012; Saltman, 2016). Data-intense systems in education can encode expectations of what a learner should be and how they should act (Decuypere, 2019) or indicate what the future holds for young people like them (Eynon, forthcoming). A concern raised by many critical scholars is the ways that digital trace data is being used in schools supports a strong measurement regime, where constant monitoring becomes the norm (Jarke and Breiter, 2019) which is likely to lead to, among other things, an acceptance of monitoring in all contexts and life stages. An important question is what kind of world, characterised by the use of digital trace data, does society want young people to be socialised into?

Data and subjectification

The role of data-driven systems in schools are often seen as a way to free up the time of teachers. This argument is not new – Skinner, one of early pioneers of teaching machines, also argued for the importance of using such machines to free up teacher time (Skinner, 1961).

Yet, automating parts of the teaching process changes how learning and teaching happens. Importantly, they "displace the dialogue between teachers and students with pre-packaged curricula" (Saltman, 2016: 113); and encourage more didactic forms of learning new things as opposed to encouraging pedagogical exchange that is important for education (Saltman, 2016: 108). The relationships between teacher and learner are an important part of education, and changing it has consequences. As Audrey Watters asked in a keynote at Berkeley City College in 2014, "What does it mean to tell our students that we're actually not going to read their papers, but we're going to scan them and a computer will analyze them instead?" (Watters, 2014: 56).

Even if data-intense systems in schools lead to students scoring more highly on certain tests, it is important also to think more about whether this is educationally desirable (Biesta, 2015). Although many academics and teachers are concerned about the extent to which such systems may have for depth of understanding, these systems may also undermine subjectification, that is, people's capacity for action (Biesta, 2015). This 'coming into presence' is only possible when others acknowledge and respond to our actions (Biesta, 2016). If students feel that their actions are not meaningfully engaged with by teachers or by other students, as could be the case in some systems, this has significant implications for education. The behavioural engineering practices such as nudging to produce learning outcomes that are characteristic of many of these systems (Knox et al., 2020) make it difficult for students to "respond in their own unique way to the learning opportunities provided by the curriculum" (Biesta, 2016: 138) in a way that facilitates a genuine interaction between teachers and students (Licenberg and Eynon, 2021).

Furthermore, these nudging characteristics of such systems might disempower and infantilise students (Selwyn, 2019). As Hartman-Caverly (2019) suggests, an individual's ability to determine what and how to deploy attention is central to the development of people as "self-sufficient learners and independent thinkers" (Hartman-Caverly, 2019: 24). The use of data methods in education such as nudging or trying to shape learner's behaviour in any way can interrupt this process, directing the attention of students (preventing intellectual freedom) and disrupting our attention and the possibilities of focusing on an issue in an in-depth way (Hartman-Caverly, 2019). This is not just about the risks of the negative impacts of data-driven systems on the possibilities for sustained engagement for meaningful learning, but the impact this may be having on our abilities to become a person who can think and act independently and to learn in a self-determined way (Hartman-Caverly, 2019).

Compounding inequalities

The influence of digital trace data on the roles of schools will vary. Thinking through the implications of using digital trace data using the framing proposed by Biesta varies significantly according to context. This chapter is written primarily from the perspective of schooling in the global north, and even within this collection of countries, experiences, practise and policies will vary significantly.

Schools are of course highly varied places. While they may experience similar pressures and constraints from policy-makers and other systems, schools vary in their practices and philosophies and the ways that they engage with and react to policy and other pressures (Lipman, 2002). Teachers and other stakeholders all have some agency in this process, they can all, "'rewrite' policies through their own actions within the restrictions imposed on them" (Lipman, 2002: 383). Nevertheless, some will have more opportunities to exert their agency than others.

Research that has explored digital inequalities in schools have long documented the differences in what some of those responses may look like (Warschauer, 2004); and it is likely the same variation will be seen with digital trace data. There are concerns that schools with less resources tend to resort to more standardised, assessment-driven tools, and in general the data tends to be more likely to dictate practice rather than be used as one of many approaches. As Zeide notes,

> students enrolled in schools which can afford to allow teachers to deviate from algorithmic recommendations or supplement automated assessment with personal evaluation will have the opportunity for accommodation based

> on individual circumstances. Less fortunate students may receive automatically differentiated instruction without the flexibility of contextualized assessment.
>
> (Zeide, 2017:169)

Indeed, outside of schools, forms of educational redlining are already emerging (e.g., the differential use of AI bots in MOOCs) with those able to pay provided with a more sophisticated human-mediated experience (Winters, et al., 2020).

This argument fits with wider concerns that draw attention to the need to focus on who is subjected to the analysis of big data systems, as those who are most marginalised and least well off are more likely to be targeted (Crawford et al., 2014: 1666). As well as perhaps being over-reliant on data systems as discussed above, the use of education digital trace data makes even more visible certain groups of students, "The 'outliers', 'under-achievers', and 'under-performers' produced by performance measures become targets of manipulation, disapproval and anxious self-scrutiny" (Espeland and Stevens, 2008:416, drawing on Hacking, 1999).

Studies have shown how students who are classified negatively in some way are then subject to different interventions. For example, in studies of primary school children in the UK children who were on a borderline pass were excluded from arts-based lessons to receive intensive phonics booster classes (Roberts-Holmes and Bradbury, 2016). Higher achievers were left to succeed on their own, a low achieving group were considered hopeless cases, and those children in the middle ground were given enough support to get them to the right level (Roberts-Holmes and Bradbury, 2016). As Spielman notes,

> It is a risk to social mobility if pupils miss out on opportunities to study subjects and gain knowledge that could be valuable in subsequent stages of education or in later life. Restricted subject choice for low-attaining pupils disproportionately affects pupils from low-income backgrounds.
>
> (Spielman, 2017: n.p.)

Yet this approach, due to the demands of league tables is happening routinely in some schools.

These trends are perhaps not new, but intensify with data-driven systems, and indeed are likely to be further compounded by the likely biases in these systems. Systems designed to facilitate school choice and integration, evaluate essay writing, detect concentration and emotion in the classroom, evaluate the effectiveness of teachers, assess students, check for cheating and map attendance are just some of a growing number of examples of systems that can favour certain groups of students over others (Crawford et al., 2019; O'Neil, 2016; Watters, 2014; Whittaker et al., 2018).

Current responses

Given the arguments made above, it would seem appropriate for a robust response to ensure the ways that digital trace data is used in schools leads to the kinds of education systems that society wants for young people. Yet, in reality the current responses both in and outside school are inadequate. This is, for the most part, because the practical and policy-orientated responses to digital trace data essentially treat such data as if it is simply a bigger or better set of data than has traditionally been collected in education systems. It is not considered as something that can lead to qualitative changes in the role of schooling, just as something that improves existing processes.

Outside education

This has implications for the kinds of policy responses to digital trace data that have been seen across the globe. For example, by framing digital trace data as a straightforwardly good thing, collecting student data becomes acceptable as more data leads to better models, and this will lead to digital trace data doing the best for the most students. Within such a frame, discussions move quickly to the legal basis to protect individuals' information or other appropriate governance practices. Thus issues of the use of digital trace data becomes a debate about legal and practical requirements rather than wider questions of values within education (Bulger, 2016; Eynon, 2013; Hakimi et al., 2021). Issues of privacy and data protection, for example, are extremely important but not sufficient.

Legal frameworks are problematic as they tend not to keep up to date with technical trends and also do not account for the wider social and educational implications of such approaches. For example, a legal analysis of US policy highlighted that legal provisions in schools focused on information practice principles, but this neglected more value orientated concerns around the use of such data such as questions of student agency or the risks of a discrimination (Regan and Jesse, 2019). Similarly, while there has been significant progress in relation to privacy and data protection (e.g., the EU's General Data Protection Regulation (GDPR), introduced in 2018) that is important in education, it does not fully account for the ways that the move towards data-driven systems in education has implications for human rights (Berendt et al., 2020). The adoption of General Comment 25 (2021) on children's rights in relation to the digital environment by the UN Committee on the Rights of the Child is a highly significant move, yet more work needs to be done to consider what kinds of values we want to promote in technology for young people in and outside of education (Livingstone, 2019).

Alongside legal responses, many companies are setting up governance structures. Yet, there are concerns that these governance structures operate as self-regulatory checklists that substitute for independent legal or regulatory oversight (Wagner, 2018). Many of these governance structures are increasingly contested, including concerns about practices of 'ethics washing' where companies promote an image of concern about ethics, while fundamental practices remain unperturbed by these public facing activities (Wagner, 2018). They also tend to focus on security, transparency, accountability and reliability of data (Hakimi et al., 2021) – which while important does not fully address the kinds of issues raised above.

Relatedly there are growing concerns that the commercial sector has too much power in determining ethical governance and regulation of the use of digital trace data and AI across all areas of social life (Benkler, 2019). As Macgilchrist (2019) notes,

> no matter how good the motives, and how pedagogically well-founded the decisions, it is a post-democratic moment when the ability to make these decisions has shifted from publicly accountable government officials, policy-makers or educators, to developers, programmers, designers and other staff in private EdTech organisations.
>
> (Macgilchrist, 2019: 83)

Within education

Schools, for the most part, have echoed this primarily instrumental response. In other words, they tend to support (in practical terms) the use of data to enhance efficiency, increase transparency, support competitiveness, evaluate performance and improve the learning experience. This fits with a wider concern around school policies in the recent times, that school policy is largely reactive instead of proactive, in that it is based on a relatively passive response to the perceived requirements of society. It is based on the assumption that there is a need to act and benefit from the changes occurring in our society, and to do so fast, rather than thinking about schools in a way that help create the society we want (Biesta, 2013).

This is likely partly due to the ways data encourages the creation of education markets that encourage "a culture of self-interest" by promoting parental and student choice (Ball, 2017: 54), thus focusing attention on individual and school needs, as opposed to wider social and educational concerns. Performativity, both of educators and of students, is a significant concern (Ball, 2017). Data encourages a situation where teachers begin to "'want' what the system needs in order to perform well" (Lyotard, 1984: 62)" (Ball, 2017: 55). For example, a study of primary schools in the UK highlighted how even though teachers were aware of the problems of all the data they were asked to collect, they tended to comply with the requests, adding more self-governance to an already challenging workload (Roberts-Holmes and Bradbury, 2016). Schools stop talking about what schools are for, and instead focus on indicators of "quality [that] are taken as a definition of quality" (Biesta, 2019: 3).

The acceptance of data as a good thing is also reflected in the coding agendas seen in many school systems. Coding is promoted as a way to guarantee success for the individual and society – but coding does not translate into the same opportunities for everyone. All young people have different levels of economic, social, cultural and political resources and this clearly influences the kinds of opportunities they are able to get from new technologies. Yet, the focus has become very narrow (on code literacy) and not on data literacy (which is what would enable people to think about data structures as well as algorithms) (Driscoll, 2012). As a society, it is important to make decisions about how society wishes young people to use and engage with data-intense systems, and how education systems are developed to support that vision.

Reconfiguring data use in schools

Digital trace data has all kinds of implications for the roles of schools that require a considered response. This requires more meaningful debate amongst all actors in the ecosystem: policy-makers, the commercial sector, academics, students and the general public. As part of this strategy, it is imperative to move away from a system where commercial interests dominate the discussion.

It is perhaps a good moment for such a shift. The COVID-19 pandemic has thrown light on the importance of schooling and the complex role schools play in society; it has also made visible the significant social, educational and technological inequalities in society. The examination controversies in the summer of 2020, such as those led by the UK government and Ofqual, made visible the significant problems of current assessment systems and raised questions about the extent to which data and algorithms should be privileged over professional judgement. At the same time, teachers and students were often given little choice but to learn and teach via digital technologies, and have found the many limitations as well as possibilities that these data-intense systems offer.

At a time when many are looking for positives in the most difficult of times, with hopes for creating the 'new normal' and 'building back better' there may be opportunities to raise debate in this area. For example, the experiences during the pandemic may well enable broader conversations about the role of school in societies and how technology and related data-driven systems should be used in education. Teachers may have a newfound respect in society and a return to a recognition of their professional status, with more power and a stronger sense of agency in discussions over the current and future use of data systems in education.

There is then, the potential of a heightened awareness of data-related issues amongst educational professionals and the public, which may enable sustained debate over what ends societies want data-driven education to achieve (Selwyn, 2019). Such dialogue is difficult to enable but is possible, particularly when facilitated at a local level with local concerns in mind. For example, drawing on the case of the Citizen Schools of Porto Alegre, Apple (2012) documents how

> a wider group of people (including local citizens, teachers, administrators, parents) [were] actively involved in decision-making – about budgets, about supporting knowledge construction that [was] centred within the interests and concerns of the local community as well as connecting to the national curriculum.
>
> (Apple 2012:138)

Despite this group of people not ordinarily "allowed" or expected to be part of the discussions to help to enable change, positive change occurred, with participatory mechanisms at the heart of the strategy (Apple, 2012:103).

Relatedly, D'Ignazio and Klein (2019) draw on feminist theory to argue for the importance of multiple actors needing to be involved in data work. Their approach recognises and values the different views of those involved or implicated in its use, with particular effort and attention to those who are most marginalised. Such an approach can allow for a range of possibilities that may result in positive social change. For example, participatory mapping initiatives that embed community storytelling, facilitating data informed conversations within and across communities, and using data to support consensus building. As the authors note,

> A data scientist is not going to save democracy, but a well-designed, data-driven, participatory process that centers the standpoints of those most marginalised, empowers participants and builds new relationships across lines of social difference? Well, that might just have a chance.
>
> (D'Ignazio and Klein, 2019, n.p.)

This could be highly promising when applied to educational settings.

One key approach of more participatory methods is participatory design. Participatory design has a long history of enabling both theoretical and practical change. Key to this is to actively design in the values and aims of education that society wishes to promote, and in doing so may help to challenge and reinvent the current commercial ecosystem (Macgilchrist et al., 2020). Relatedly, authors have called for an 'ethics by design' approach in educational institutions that use data analytics where ethical considerations and their renegotiation are at the heart of the design of data-based education systems and involve key stakeholders at all stages (Gray and Boling, 2016). One option, for example, may be to use the framing proposed by Biesta (2015) to help negotiate and determine those values as part of a design strategy.

It is important to be mindful that participatory design, as with all participatory methods are not straightforward. Technologies can still be designed in ways that are problematic or reinforce inequalities and there is nothing to say that users are somehow expert in what is needed (Berg, 1998). Such techniques cannot be considered a straightforward solution, but they are likely to enable better theorisation, along with a richer understanding of key issues, and enable more conversations around data and education to occur. Indeed, there are some excellent examples in this book.

A second important area is ensuring students have control and understanding of their data. As Audrey Watters argued at a keynote in 2014, when talking about the metaphors and implications of data mining, "I want to encourage the building of technologies that see students' lives and learning not as a resource to be extracted but as something they themselves can control and cultivate" (Watters, 2014: 105). This is not just about individuals being more aware of and having more control over their data usage, but instead being part of something that enables them to create a new kind of public engagement, discourse and change around the ways that data in education are envisaged and used.

Similar to Couldry's (2003) discussion on the need to shift static discussions of a digital divide towards a discussion about how the Internet can be an active space that contributes to democratic life, it is not just whether people know how to access and use their data but if they can "effectively represent their material interests" (Couldry, 2003: 94). The Internet, or indeed digital trace data, is not just something that exists and can be black-boxed, but "should remain, open to deliberative intervention and ethical inquiry" (Couldry, 2003: 90).

Education is of course a central part of this, and connects closely with calls for digital data literacy, where what is needed is a, "deliberately political model of digital literacy in which complex and detailed understandings of discourse, ideology and power in the digital context are scaffolded" (Pangrazio, 2016: 170). An important part of learning this is through the process of design, and enabling young people to produce their own artefacts (Pangrazio, 2016; Winters et al., 2020). Precisely how this can be achieved in multiple contexts is documented in other chapters throughout this book.

Young people are, of course, not the only stakeholders who need to develop understandings about digital trace data. Teachers and education stakeholders also need to learn about what data and AI is good for and what it is not so good at (Winters et al., 2020), but it is an important start and may help to reconfigure the existing direction of inquiry.

Despite the possibilities of using such approaches to reconfigure the status quo, it is important to note that schools do not operate in a vacuum. As Reay notes, "educational systems are only as good as the societies they emerge out of [...] capitalist, neoliberal societies beget capitalist neoliberal educational systems" (Reay, 2011:2). While data are at the heart of the business model of many other mainstream platforms that young people may use for learning and everyday life (Zuboff, 2019) there are significant implications for education (Hakimi et al., 2021). As Hartman-Caverly notes, "learning analytics is but one node in a broader network of surveillance capitalism, in which the power to accumulate information harvested from human behavior creates control over others' lived realities West, 2019; Zuboff, 2015" (Hartman-Caverly, 2019:40). Given the current direction of surveillance capitalism, there are important questions about how we educate for democracy in such a context (Biesta, 2019, Sefton-Green, 2020).

Conclusion: challenging the status quo

In summary, data is becoming a defining feature of our education system. Importantly, digital trace data or big data needs to be understood both technically but also socially and culturally. At present digital trace data is being framed in a particular way that promotes the marketisation of schooling and as a way to solve all educational changes. Yet who benefits most from this framing are primarily those from the commercial sector, not from education.

Using Biesta's (2015) framing to critically examine the roles of schools highlighted how the use of digital trace data has multiple implications for what schools do. Such changes can be seen at all levels, from the experiences of individual learners to whole school systems – and is likely to further disadvantage those who are already disadvantaged.

If society wishes to continue using such data within schools it is important to reconfigure the ways it is being conceptualised, designed, and implemented and this requires changes across all actors in the ecosystem. Schools should not just reflect society, they can also change it (Apple, 2012). Rather than merely reproducing and reinforce the dominant ideologies of those in power, there are possibilities for schools to instead question those patterns or indeed, facilitate a change in wider society (Apple, 2012). This can be achieved through various approaches including participatory methods and changes in curriculum from moving from a focus on coding to data literacy.

For centuries, there has been debate about the most appropriate relationship between school and society. Are schools simply there to meet the needs and demands of society, or should they be viewed as institutions slightly set apart from society? The first perspective emphasises that schools provide whatever society needs at any given point in history, in the latter; schools are places of reflection and action that may challenge and change the status quo (Biesta, 2019). Biesta notes that both are likely necessary for any democratic society, yet that at present the balance between these two positions has been lost: the dominant position is that schools should service the needs of society.

The current faith in data and data solutionism, education policies driven by the data economy, and educational tools based on a philosophy of extracting economic value from the collection and processing of data resulting from digital technologies and the people who use them, where commercial interests are prioritised (Selwyn, 2019), does not provide many opportunities for change. Yet it is possible. The other chapters in this book will discuss in more depth how this could be achieved, and how schools can challenge the current status quo.

References

Apple, M.W. 2012. *Can Education Change Society?* London: Routledge.

Ball, S.J. 2017. *The Education Debate.* Bristol: Policy Press.

Benkler, Y., 2019. Don't let industry write the rules for AI. *Nature*, 569(7754), pp. 161–162.

Berendt, B., Littlejohn, A. and Blakemore, M., 2020. AI in education: Learner choice and fundamental rights. *Learning, Media and Technology*, 45(3), pp. 312–324.

Berg, M. 1998. The politics of technology: On bringing social theory into technological design. *Science, technology & human values*, 23(4), pp. 456–490.

Berry, D. M. (2011). The computational turn: Thinking about the digital humanities. *Culture Machine*, 12. Available at: http://www.culturemachine.net/index.php/cm/article/view/440/470

Biesta, G. 2013. Responsive or Responsible? Democratic Education for the Global Networked Society. *Policy Futures in Education*, 11 (6), pp. 733–744.

Biesta, G. 2015. *Good Education in an Age of Measurement: Ethics, Politics, Democracy.* London: Routledge.

Biesta, G. 2016. *Beyond Learning: Democratic Education for a Human Future.* London: Taylor & Francis Group.

Biesta, G. 2019. What kind of society does the school need? Redefining the democratic work of education in impatient times. *Studies in the Philosophy of Education*, 38 (6), pp. 657–668.

Boellstorff, T. 2013. Making big data, in theory. *First Monday*, 18(10). doi:10.5210/fm.v18i10.4869.

Borgman, C. 2015. *Big Data, Little Data, No Data*. Cambridge, MA: The MIT Press.

Bulger, M. (2016). Personalized learning: The conversations we're not having. *Data and Society*, 22(1) pp. 1–29.

Couldry, N. 2003. Digital divide or discursive design? On the emerging ethics of information space. *Ethics and Information Technology*, 5(2), pp. 89–97.

Crawford, K., Miltner, K. and Gray, M. L. 2014. Critiquing big data: Politics, ethics, epistemology. *International Journal of Communication* 8, pp. 1663–1672.

Crawford, K. et al. 2019. AI Now 2019 Report. New York: AI Now Institute, 2019, https://ainowinstitute.org/AI_Now_2019_Report.html.

Davies, H. C., Eynon, R. and Salveson, C. 2020. The mobilisation of AI in education: A Bourdieusean field analysis. *Sociology*, 55(3), pp. 539–560.

Davies, C. and Eynon, R. 2013. *Teenagers and Technology*. London: Routledge.

Decuypere, M., 2019. Researching educational apps: Ecologies, technologies, subjectivities and learning regimes. *Learning, Media and Technology*, 44(4), pp. 414–429.

D'Ignazio, C., and Klein, L. 2019. *Data Feminism*. Cambridge: MIT Press.

Driscoll, K. 2012. From punched cards to "Big Data": A social history of database populism. *Communication*, 1(4), pp. 1–33.

Espeland, W. N. and Stevens, M. L. 2008. A sociology of quantification. *European Journal of Sociology*, 49(3), pp. 401–436.

Eynon, R. 2013. The rise of Big Data: What does it mean for education, technology, and media research?. *Learning, Media and Technology*, 38(3), pp. 237–240.

Fischer, C., Parados, Z. and Baker, R. 2020. Mining big data in education: Affordances and challenges review of educational research. *Review of Research in Education*, 44(1), pp. 130–160.

Gray, C. M. and Boling, E. 2016. Inscribing ethics and values in designs for learning: A problematic. *Educational Technology Research and Development*, 64, pp. 969–1001.

Hacking, I. (1999). *The Social Construction of What?* Cambridge, MA: Harvard University Press.

Hakimi, L., Eynon, R., and Murphy, V. (2021). The ethics of using digital trace data in education: A thematic review of the research landscape. *Review of Educational Research*. doi:10.3102/00346543211020116.

Hartman-Caverly, S., 2019. Human nature is not a machine: On liberty, attention engineering and learning analytics. *Library Trends*, 68(1), pp. 24–53.

Jarke, J. and Breiter, A. 2019. Editorial: The datafication of education. *Learning, Media and Technology*, 44(1), pp. 1–6.

Kitchin, R. 2014. Big Data, new epistemologies and paradigm shifts. *Big Data & Society*, 1(1), pp. 1–14.

Knox, J., Williamson, B. and Bayne, S. (2020). Machine behaviourism: future visions of 'learnification'and 'datafication'across humans and digital technologies. *Learning, Media and Technology*, 45(1), pp. 31–45.

Laney, D. (2001). 3D Data Management: Controlling data volume, velocity and variety. META Research Note, 6. Gartner Report. Gartner.

Licenberg, J. and Eynon, R. (2021) Crafting worldly spaces: The role of the educator in shaping student agency in minecraft. *Review of Education, Pedagogy, and Cultural Studies*. doi:10.1080/10714413.2021.1970465.

Lipman, P., 2002. Making the global city, making inequality: The political economy and cultural politics of Chicago school policy. *American Educational Research Journal*, 39(2), pp. 379–419.

Livingstone, S. (2019). Claiming rights and righting wrongs: Children's rights in the digital environment. In: Donohue, Chip, (ed.) *Exploring Key Issues in Early Childhood and Technology: Evolving Perspectives and Innovative Approaches.* New York, NY: Routledge, pp. 42–48. ISBN 9781138313798.

Luckin, R., Holmes, W., Griffiths, M. and Forcier, L. B. 2016. *Intelligence Unleashed. An Argument for AI in Education.* London: Pearson.

Luke, A. (2009). Critical realism, policy and educational research. In Ercikan, K. and Roth, W.M. (eds.), *Generalizing from Educational Research* (pp. 173–200). New York: Routledge.

Lyotard, J.-F. (1984). *The Postmodern Condition: A Report on Knowledge.* Manchester: Manchester University Press.

Macgilchrist, F. 2019. Cruel optimism in EdTech: when the digital data practices of educational technology providers inadvertently hinder educational equity. *Learning, Media & Technology,* 44(1), pp. 77–86.

Macgilchrist, F. Allert, H. and Bruch, A. 2020. Students and society in the 2020s: Three future 'histories' of education and technology. *Learning, Media and Technology,* 45(1), pp. 76–89.

Markham, A. M. 2013. Undermining 'data': A critical examination of a core term of scientific inquiry. *First Monday,* 18(10). doi:10.5210/fm.v18i10.4868.

O'Neil, C., 2016. *Weapons of Math Destruction: How Big Data Increases Inequality and Threatens Democracy.* London: Allen Lane.

Ozga, J. 2016. Trust in numbers? Digital Education Governance and the inspection process. *European Educational Research Journal,* 15(1), pp. 69–81.

Pangrazio, L. 2016. Reconceptualising critical digital literacy. *Discourse: Studies in the Cultural Politics of Education,* 37(2), pp. 163–174.

Porter, T. 1994. Making things quantitative. *Science in Context,* 7(3), pp. 389–407.

Powers, J. M., Fischman, G. E. and Berliner, D. C. 2016. Making the visible invisible: Wilful ig norance of poverty and social inequalities in the research-policy nexus. *Review of Research in Education,* 40(1), pp. 744–776.

Reay, D. 2011. Schooling for democracy: A common school and a common university? A response to "schooling for democracy". *Democracy and Education,* 19(1), Article 6. Available at: https://democracyeducationjournal.org/home/vol19/iss1/6.

Regan, P. M. and Jesse, J., 2019. Ethical challenges of EdTech, big data and personalized learning: twenty-first century student sorting and tracking. *Ethics and Information Technology,* 21(3), pp. 167–179.

Roberts-Holmes, G. and Bradbury, A. 2016. The datafication of early years education and its impact upon pedagogy. *Improving Schools.* doi:10.1177/1365480216651519

Robins, K. and Webster, F. 1989. *The Technical Fix: Education, Computers, and Industry.* London: Macmillan.

Romero, C. and Ventura, S., 2020. Educational data mining and learning analytics: An updated survey. *Wiley Interdisciplinary Reviews: Data Mining and Knowledge Discovery,* 10(3), p. e1355.

Saltman, K. J. 2016. Corporate schooling meets corporate media: Standards, testing, and technophilia. *Review of Education, Pedagogy, and Cultural Studies* 38 (2), pp. 105–123.

Sefton-Green, J., 2020. Educating for democracy in the digital age. In Noblit, George W. and Neikirk, Joseph R. (eds.), *Oxford Research Encyclopedia of Education.* Oxford: Oxford University Press. doi:10.1093/acrefore/9780190264093.013.1005.

Selwyn, N., 2019. What's the problem with learning analytics? *Journal of Learning Analytics,* 6(3), pp. 11–19.

Spielman, A. (2017) *HMCI's commentary: October 2017.* Available at, https://www.gov.uk/government/speeches/hmcis-commentary-october-2017.

Skinner, B. F., 1961. Teaching machines. *Scientific American,* 205(5), pp. 90–106.

Wagner, B., 2018. Ethics as an escape from regulation: From "Ethics-Washing" to ethics-shopping? In E. Bayamlioglu, I. Baraliuc, L. Janssens & M. Hildebrandt (Ed.), *Being Profiled: Cogitas Ergo Sum: Cogitas Ergo Sum: 10 Years of Profiling the European Citizen* (pp. 84–89). Amsterdam: Amsterdam University Press. doi:10.1515/9789048550180-016.

Watters, A., 2014. *The Monsters of Education Technology*. Ebook.

Warschauer, M. (2004). *Technology and Social Inclusion: Rethinking the Digital Divide*. Cambridge: MIT Press.

Webster, F., 2014. *Theories of the Information Society*. London: Routledge.

Whittaker, M., Crawford, K., Dobbe, R., Fried, G., Kaziunas, E., Mathur, V. et al. (2018). AI now report 2018. New York: AI Now Institute at New York University.

Williamson, B. 2017. Who owns educational theory? Big data, algorithms and the expert power of education data science. *E-Learning and Digital Media*, 14(3), pp. 105–122.

Williamson, B., 2021. Making markets through digital platforms: Pearson, edu-business, and the (e) valuation of higher education. *Critical Studies in Education*, 62(1), pp. 50–66.

Winters, N., Eynon, R., Geniets, A., Robson, J. and Kahn, K., 2020. Can we avoid digital structural violence in future learning systems? Learning. *Media and Technology*, 45(1), pp. 17–30.

Zeide, E., 2017. The structural consequences of big data-driven education. *Big Data*, 5(2), pp. 164–172.

Zuboff, S. 2019. *The Age of Surveillance Capitalism: The Fight for a Human Future at the New Frontier of Power*. London: Profile Books.

3

TURN OFF YOUR CAMERA AND TURN ON YOUR PRIVACY

A case study about Zoom and digital education in South American countries

Cristobal Cobo and Pablo Rivera Vargas

Introduction

In the 21st century, education can hardly be decoupled from digital technologies (EdTech). Regardless of the income level of the country, teaching and learning policies are closely intertwined the with the influence of digital technologies in education policies. "EdTech is growing at 16.3% and will grow by 2.5 times from 2019 to 2025, reaching $404 billion in total global expenditure" (Holon IQ, 2020). There are reasons to believe that EdTech will grow, especially after the massive adoption of remote or blended learning policies generated during COVID-19 around the world.

If that were the case, policymakers and the EdTech community will need to carefully reflect on the responsibilities derived from an intensive digitisation of education systems. On the one hand, EdTech could reduce the cost of content delivery and potentially make it possible to reach those learners who are harder to reach. On the other hand, there are transactional costs that come with the adoption of the EdTech policies. Policy makers in charge of implementing EdTech policies will need to address some of the associated challenges such as "different forms of exclusion", "commodification of education" (Schwartzman, 2013), or "datafication of education" (Jarke and Breiter, 2019). In this chapter, we will explore the latter.

The COVID-19 outbreak activated an explosive use of digital technologies to enable remote learning, making Zoom (the global video conference provider) one of the most popular educational technologies. Zoom has faced several global concerns (e.g., private information shared with third parties, adoption of dubious end-to-end encryption processes, exposing users to unauthorised bullying calls).

In this study we explore how different South American countries govern the use of commercial platforms in education systems and to what extent they alert their citizens from the potential risks. The analysis focused on public education

DOI: 10.4324/9781003136842-3

institutions from eight countries (Argentina, Brazil, Colombia, Chile, Ecuador, Paraguay, Peru, and Uruguay), and how they reacted (or not) to protect the citizenship and especially the educational community privacy at the beginning of the pandemic in 2020.

The questions that guided this work were: How public institutions of South America reacted when the privacy risks of Zoom were made known? Where did they put their emphasis? What lessons can be learned? From these questions, we have organised the chapter into four sections: (1) General background; (2) Methodology; (3) Results; and (4) Discussion and conclusions.

General background

The problem derived from data-intensive education

Almost every action or transaction that takes place online by an education user (e.g., student or teacher) leaves traces. These can be monitored and used (today or in the future) by someone (known or unknown) who can be a subject (person) or an object (algorithm). The more transactions an education beneficiary generates online, the more information can be monitored later.

Often, monitoring actions are presented by digital or platform providers as something that will benefit the user by "improving the service" or by enhancing "the quality of the experience online". However, that is not necessarily the case. It is not always clear how sophisticated algorithms use users' information, how this information is managed, exploited or commercialised. This lack of transparency produces asymmetries of information between the user and the digital service provider (Cobo, 2019) raising concerns and conflicts of interest associated with data protection and privacy.

In the educational area, when planning guidance and strategies for mitigating privacy-related risks, EdTech policy makers will face the challenge of understanding the business model of (most) digital providers. The data monetisation model has its origins in the early days of the big digital corporations (Mack, 2019). Back then, data was used mostly to improve the quality of the user experience. However, the lack of a clear revenue model pressured these corporations to find ways to exploit data from users for "other" purposes (e.g., targeted advertising or related) (Digital Future Society, 2019).

Technology developers and software companies have increasingly invested time and resources in the education sector (HolonIQ, 2020). However, just like in other sectors the interest in extracting data from users has become a central component of the revenue model (Reidenberg and Schaub, 2018; Komljenovic, 2020). Given that EdTech has gained prominence in education policies, the challenges associated with the data privacy of students and teachers are acute. Privacy is not only a concern only for legal experts. Quite the opposite, privacy (and the protection of personal data in general) shall be understood as a core component when weighing the pros and cons of EdTech.

The importance of privacy

Privacy is a multidimensional and contextual phenomenon. According to Brewczyńska (2020), it has three layers: (a) the data provided by the user such as name, mail or phone; (b) the data that the user leaves while using digital platforms, cookies or browser history and (c) the data interpreted by algorithms from a and b, which might be true or might not. Privacy is transactional; it is usually understood as an individual good that needs to be "sacrificed" for a greater benefit, but when lost is hard to get it back (Snowden, 2019). The concept of privacy is often confused with the necessity to hide something (Bernal, 2020). Many argue "I don't have anything to hide, therefore I don't need to worry about my privacy". Particularly in digital environments privacy is not valuable simply because users (students) can protect their intimacy, identity, personal protection, but also because the violation of privacy can lead to different abuses of power (e.g., control, surveillance, manipulation, discrimination, etc.).

The larger the volume of data exchanged digitally, the more sophisticated the systems need to be to gather and make sense of the data, therefore the more difficult it will be for non-expert users to understand how to protect their personal data (Raffaghelli, 2020). The sophistication of these systems demands from education policy makers a whole new data governance agenda. The lack of appropriate technical and political measures could expand the technological and cognitive divide between those who create those systems and the large volume of users (e.g., students, teachers) who have a limited understanding of how those systems work (Williamson and Piattoeva, 2019). In order to reduce these information asymmetries (and the associated risks) education ministries and related agencies need to prepare their technical teams and regularly revise the relevance and the effectiveness of their actions.

The underlying business models

The business model adopted by most platform providers is free of payment (non-monetary transaction involved) which is usually misunderstood as "free service" (Pardo, Waliño-Guerrero and San Martín, 2018). The digital services are, in most cases, supported by Terms and Agreements where the users consent to give access to their personal data (as well as the possibility of trading that data with other parties). Today there is a wealth of examples where "personal data" has become a "new" form of currency to pay for the online services. This "new" currency gives power to those "data brokers" (often Big Tech), which are granted access to individuals' data. Unfortunately, this is still not fully understood nor sufficiently regulated in defence of the education community who ignore (or prefer to not think about) the transactional cost of these "free" digital services (Burch and Miglani, 2018; Sancho-Gil, Rivera-Vargas and Miño-Puigcercós, 2020).

A naive understanding of the business model adopted by internet companies has significantly eroded people's privacy (Snowden, 2019). However, not only have individuals failed to understand this digital reality, but also multinational regulatory entities, as well as national legislative institutions, have not fully addressed the associated risks of a data-intensive economy until very recently. EU General Data Protection Regulation became law in May 2018 (van Veen, 2018).

The invasion of privacy has been used as a tool for manipulation. Gathering large volumes of information enables companies to elaborate a complex profile of EdTech users, feeding the existing records about individual's behaviour, exchange, or performance (Cho, et al., 2020). When this profiling is done not individually but collectively (Véliz, 2020) (analysing hundreds of thousands if not millions of people simultaneously) then it is possible to enable massive personalisation at scale (every message includes an advanced level of personalisation customising the content adjusted to the individual and their situation or context).

Different data protection experts (Brewczyńska, 2020; Véliz, 2020; Snowden, 2019) argue that privacy should be an individual as well as a collective principle, which among others, should be addressed and protected by public institutions. In the context of education, it is unrealistic to expect that every student or teacher has to personally protect their privacy against "Big Tech", which have developed sophisticated technical and legal systems to hoover up data from their users (Morozov, 2018). The misuse of personal information can have the gravest consequences. For instance, when citizens entrust personal information to companies like Facebook or Google and that trust is broken, there is an abuse of a dominant position. Data collection has become an economic factor, which in turn makes data protection a potential drag on economic growth (Cobo, 2019). What would it take to increase the level of data protection in education? With millions of students daily using the online platforms from "Big Tech", it is critical to increasing the transparency of how personal data is being used, making these organisations accountable every time there is a misuse of users' personal data (Reidenberg and Schaub, 2018). On the other side, additional protection measures need to be taken from the education systems. For instance, education and public institutions should always take pre-emptive measures to educate, guide, and protect individuals before confronting these risks (Pangrazio and Selwyn, 2020).

The importance of the governance of digital platforms in educational institutions

The governance of digital education is an area of increasing relevance, as illustrated in the case of Zoom. Educational systems are expected to develop advanced capacities such as comprehend, monitor, and manage increasingly sophisticated digital systems, protecting institutions without compromising individuals' rights or their personal data. The governance of digital platforms requires the articulation of sociotechnical factors, such as adopting key principles, accountability protocols and the technical infrastructure to mitigate potential risks. Equally relevant is guiding education users with the technical awareness, training and support needed to effectively integrate technology without risking the integrity and data protection of the member of this community (Williamson, 2016; Selwyn, 2016).

All of the challenges associated with the intensive use of personal data raise alarms for EdTech policies adopted in different countries. While the technology itself evolves at an exponential rate, the capacity of the institutions in charge of regulating and facilitating education at national levels tend to struggle to catch up with these rapid transformations (the situation worsened when proactive measures are needed) (Regan and Jesse, 2019).

Although EdTech could enhance learning opportunities, new institutional capacities will be required when thinking about data protection and privacy matters. Education organisations must adopt actions to minimise the risks and to protect confidential information and the identities of their users. What data is collected? Where and how is it collected? Who has access to the data and for what purposes? These among other questions need to be explored. As presented below, COVID-19 made it more critical than ever to react in a timely and effective fashion.

COVID-19 situation: a (complex) window of opportunity

Given that COVID-19 spreads extraordinarily quickly by air, many of the world's governments decreed the home confinement of all people who did not have to perform essential activities (Choe and Choi, 2020). During the confinement period, "educational activities would be maintained through distance and 'online' modalities, whenever possible". Schools and Universities closed around the world almost overnight, affecting 1,576 million students in 191 countries, 91% of all students around the world (Delgado, Illanes, and Lucio-Villegas, 2020; UNESCO, 2020).

These closures forced many to transition to technology-mediated distance education. Big Tech took advantage of this "window of opportunity" and made different digital platforms available to educational institutions and communities, supposedly in an "altruistic and generous" way. COVID-19 has been a golden opportunity to expand the relevance of Big Tech. COVID-19 exponentially increased Internet use and dependence by up to 60% in OECD countries (Encinas-Martín, 2020), because most people were forced to work, study, perform leisure activities (watching movies or series, using apps to play sports at home, etc.) and communicate through video calls from their homes.

Just when Big Tech was going through a phase of manifest global questioning and in the crosshairs of the governments of many countries for their monopolistic practices (Tracy, 2020) tax fraud (Fernández, 2019), misuse of collected data (Menéndez, 2020), and "abuse of dominance position" (García, 2017), the COVID-19 pandemic broke, and Big Tech took advantage of this "window of opportunity" (Sádaba, 2016), making their different tools and digital platforms available to citizens self-promoting their "altruistic" services exemplified by:

- Alphabet Inc. (G-Suite) signed agreements with hundreds of governments to provide the educational community with its platforms and virtual tools. (Gallegos Cárdenas, 2020).
- Microsoft (Microsoft Teams educational platform), offered free access to organisations and schools around the world. (Espinoza, 2020).

- Coursera (Coursera Together), made its training offer available to all universities for free. (Coursera, 2020).
- TikTok generated the #LearnOnTikTok initiative to promote online education during confinement. (Hutchinson, 2020).
- Zoom offered a subscription and free use service (for 40 minutes per session) and became the most used app in the world for conducting video conferences (BBC News Mundo 2020), although its success was shaken by substantial privacy and security concerns.

(García, 2020)

The Zoom phenomenon

In response to the pandemic situation, schools and education institutions quickly decided to activate digital tools for remote learning, videoconferences in general and Zoom, in particular, became a critical enabler during this health crisis. Unfortunately, remote schooling put children's privacy at the mercy of EdTech companies (cfr. Morrison, 2020).

At the end of December 2019, the number of daily participants using the meeting services held on the Zoom platform, both free and by subscription, was approximately 10 million. In April of 2020, Zoom reached more than 300 million daily participants (cfr. Morrison, 2020).

Early in the crisis, different voices criticised the video conference company for falling short of providing sufficient measures to secure its privacy controls or cybersecurity. "Zoombombing" was a cause of major concern, subjecting learners to the unsolicited appearance of hackers in the middle of their classes (Secara, 2020). Zoom claimed to offer end-to-end encryption but that was a matter of broad dispute (Kari, 2020) as the company also sent user data to Facebook and LinkedIn. For instance, the American Federal Trade Commission (Federal Trade Commission, 2020) in the US expressed their concerns in this regard, but these were not addressed by the company until late 2020 (O'Flaherty, 2020).

Methodology

Methodological design

For the purposes of this chapter, qualitative research was carried out in two phases. In the first phase, considering the characteristics of the problem addressed, the authors adopted a position close to the postulates of virtual (Hine, 2004) and multi-sited (Falzon, 2009) ethnography. This perspective allowed us to examine the digital idiosyncrasy of the countries studied, analysing more than one space at the same time, and simultaneously identify the connections between them (Morozov, 2018; Miño et al., 2019). Using virtual ethnography, we identified social and educational policies, along with models of digital governance and explored the emphasis that different states took when promoting digital citizenship, privacy, and security of internet browsing (Raffaghelli, 2020).

In the second phase, once the relevant cases were defined, a thorough analysis and revision were conducted following the pragmatic discourse analysis method (Wetherell and Potter, 1998; Fardella et al., 2017). This position allowed us to explore and analyse in greater depth the three cases that were most significant considering the abundant information they contained and the social impact they generated in their respective countries.

Sample and information gathering

During the investigative phase, an intentional sampling was carried out, including ten countries from South America (Spanish- and Portuguese-speaking). However, at the beginning of the exploratory work, the selection of countries was reduced to eight, as we could not easily access online information from public institutions in Venezuela and Bolivia. The final list of countries included was Argentina, Brazil, Bolivia, Colombia, Chile, Ecuador, Paraguay, Peru and Uruguay.

Considering this final number of countries included in the study, and to favour the analytical exercise, a grouping of countries into three clusters was generated based on four indicators of substantial regional and global relevance.

- The UNDP Human Development Index (2019).
- The report on "Individuals using the Internet", created by the World Bank (2019).
- Households with a personal computer (%) from ITU (2018).
- The Basic digital conditions for online learning in South America (2020) of the Inter-American Development Bank.

It is important to note that these indicators are not the only ones available to analyse the development conditions and access to technology of the population; nor is this an exercise based on an exhaustive search for relevant indicators. However, for this study we have chosen these four indicators to generate a cluster of countries allowing us to analyse general characteristics and patterns identified within these South American countries.

As indicated in Table 3.1, the conditions between these eight countries vary according to human development, internet access, access to technological devices and social conditions for online learning.

According to the differences expressed in these indicators, we have grouped the countries into three clusters. Each of these is based on the social conditions in which each territory is situated, that is to say: advanced, emerging and frontier.

- Group 1: Countries with *advanced* social development conditions and technological access (Argentina, Chile and Uruguay)
- Group 2: Countries with *emerging* social development and technological access conditions (Brazil, Colombia and Peru)
- Group 3: Countries with *frontier* social development conditions and technological access (Ecuador and Paraguay).

TABLE 3.1 Human development and access to the digital society

Country	*IDH Latam (2019) (0.8 to top, Very High HD; 0.7 to 0.79, High HD)*	*Individuals using the Internet(% of population) (2019) (WordBank, 2019)*	*Households w/personal computer (%) (2018) Latam*	*Basic digital conditions for online learning in South America (IDB, 2020) (low, low medium, high-medium, high)*
Argentina	0.845	74	64	High-medium
Brazil	0.765	67	46	High-medium
Colombia	0.767	65	44	High-medium
Chile	0.851	82	60	High-medium
Ecuador	0.759	57	41	Low medium
Paraguay	0.728	69	25	Low
Peru	0.777	60	29	High-medium
Uruguay	0.817	77	71	High

Field of work

A total of 48 websites corresponding to official institutions from these eight countries were explored. This exercise led us initially to identify ten relevant documents that referred to security recommendations, alerts or protocols for conducting videoconferences in education of these countries (Table 3.2).

We identified the website of the ministries of education in each of the selected countries. Once there, and based on three attempts, we wrote in the "search engine" of each website some keywords that helped us to locate and focus the search (Table 3.3)[1].

In those websites that had the option of carrying out an "advanced search", we narrowed down the searches in two periods of time as described in Table 3.4.

We wanted to group the information in such a way that it would allow us to see the preparation before the pandemic, and the subsequent reaction by these institutions once the Zoom security breach alert came to light.

In those cases where there was no substantive information derived from the websites of the ministries of education and related entities, we went to the Google search engine and set in the "advanced search" option some parameters to collect the information required on the websites about other relevant Secretaries of State or public institutions in these countries. The parameters that we established in the "advanced search" option were:

- Google[2]/ Preferences / advanced search / Region (Country), Language (Spanish)
- Google/ Tools / Date / Custom Interval

Once these parameters were established, the concepts described in Table 3.2 were included, as a result obtaining news, communications and documents referring to the risks of violation of privacy and misuse of personal data in videoconferences.

TABLE 3.2 General information collected

	Type of document	*Title*	*Institution*	*Date of publication*	*Link*
Argentina	Recommendation	Cuidados al momento de armar una clase virtual con videoconferencia	Educ.ar, Ministry of Education	14/10/2020	https://bit.ly/3qHBXrm
	Recommendation	¿Cómo cuido la privacidad y seguridad en una videollamada?	Programme "con vos en la web" Secretary of State Justice	6/7/2020	https://bit.ly/2JELmiL
Brazil	Security alert	Ministério da Justiça e Segurança Pública notifica o aplicativo Zoom	Secretary of State for Justice and Public Security	08/04/2020	https://bit.ly/3n2AtWl
Colombia	Security alert	Resolución n74519	Secretary of State for Commerce, Industry and Tourism,	23/11/2020	https://bit.ly/2LpP3tb
Chile	Recommendation	Informe sobre aplicaciones para videoconferencia	Secretary of State of the Interior and Public Security	5/4/2020	https://bit.ly/33ZKzzH
Ecuador	Recommendation	instructivo para la implementación de educación abierta	Secretary of State Education	30/7/2020	https://bit.ly/2W0FqCU
	Recommendation	Cuidado con el ¡Secuestro! de Videoconferencias	Secretary of state for Telecommunications and the Information Society	21/10/2020	https://bit.ly/36YsCU8
Paraguay	Recommendation	Recomendaciones para mantenerte seguro con Zoom	Secretary of state for Information and Communication Technologies	13/4/2020	https://bit.ly/36XMd6T
Peru	Security alert	Alerta integrada de seguridad digital n001	Secretariat of digital government	29/03/2020	https://bit.ly/2VYZFRM
Uruguay	Recommendation	Recomendaciones para la adopción de software de videoconferencias	Agency for Electronic Government and the Information and Knowledge Society. Presidency of the Republic	30/04/2020	https://bit.ly/33Szm3Z

TABLE 3.3 Keywords

	Attempt 1	*Attempt 2*	*Attempt 3*
Keywords	Videollamada, videoconferencia, privacidad, datos personales, Zoom, vulnerabilidad, recomendaciones, seguridad	Google meet, Google Suite, Microsoft team, educación online, condiciones de uso, educación remota,	Protocolo, navegación segura, COVID-19, pandemia, alerta, confinamiento

TABLE 3.4 Considered periods

Period	*Data*
Pre-pandemic	January 1, 2015 to March 20, 2020
Pandemic	From March 21 to July 7, 2020

In the second phase, and consistent with the process, a selection of eight relevant documents from each of the eight countries (Table 3.8) was reached: Argentina, Brazil, Colombia, Chile, Ecuador, Paraguay, Peru and Uruguay. The documents were divided into two types: (1) Security alerts; (2) Recommendations for conducting secure videoconferences. The choice of the sample was carried out under the criterion of representativeness of the qualitative sampling in documentary analysis developed by Ibáñez and Iñiguez (1996), where the texts are representative as long as they present and demonstrate the discourses to be studied, and are at the same time, the usual discourses of the context of the study (Iñiguez and Antaki, 1994).

According to Prior (2008), the documents are not only produced by the institutions, but also they simultaneously produce orders and particular actions. For this reason, we worked with manifestly prescriptive documents (instructions, forms, resolutions, protocols, recommendations).

The pragmatic discourse analysis method (Edwards and Potter, 1992; Potter, 1998; Wetherell and Potter, 1998) assumes texts as discursive practices that develop a social action in their environment. The analysis then focuses on the study of these actions and the development of hypotheses about the purposes of the text (Wetherell and Potter, 1998). Successive readings, considered the questions that guided the development of this chapter: How did the public institutions of South America react when the privacy risks of Zoom were made known? Where did they put their resources? What lessons can be learned? The texts were coded, and those codes were grouped into an emerging category entitled: "Security incidents; alerts, locations and recommendations for use". Within this category, the speech was organised based on the main security incidents identified in the analysis of the eight documents (Table 3.9).

Finally, the reference to the analysed documents is made based on the acronym of each of these that appear in Table 3.5.

TABLE 3.5 Documents selected for analysis

	Type of document	*Title*	*Institution*	*Date of publication*	*Link*	*Acronym*
Argentina	Recommendation	Cuidados al momento de armar una clase virtual con videoconferencia	Educ.ar, Ministry of Education	06/07/2020	https://bit.ly/3eQoj1s	EDUCAR-Argentina
Brazil	Security alert	Ministério da Justiça e Segurança Pública notifica o aplicativo Zoom	Secretary of State for Justice and Public Security	08/04/2020	https://bit.ly/3n2AtW1	MJPS-Brazil
Colombia	Security alert / Recommendation	Resolución n74519	Secretary of State for Commerce, Industry and Tourism,	14/05/2020	https://bit.ly/2LpP3tb	MINCIT-Colombia
Chile	Security alert/ Recommendation	Informe sobre aplicaciones para videoconferencia	Secretary of State of the Interior and Public Security	5/4/2020	https://bit.ly/33ZKzzH	CSIRT-Chile
Ecuador	Recommendation	Cuidado con el ¡Secuestro! de Videoconferencias	Secretary of state for Telecommunications and the Information Society	21/05/2020	https://bit.ly/36YsCU8	TIS-Ecuador
Paraguay	Security alert/ Recommendation	Recomendaciones para mantenerte seguro con Zoom	Secretary of state for Information and Communication Technologies	13/4/2020	https://bit.ly/36XMd6T	MITIC-Paraguay
Peru	Security alert / Recommendation	Alerta integrada de seguridad digital n001	Secretariat of digital government	29/03/2020	https://bit.ly/2VYZFRM	SGD-PERU
Uruguay	Security alert/ Recommendation	Recomendaciones para la adopción de software de videoconferencias	Agency for Electronic Government and the Information and Knowledge Society. Presidency of the Republic	30/04/2020	https://bit.ly/33Szm3Z	CERT-Uruguay

Results

This section presents the main results of this study. In the first part, we present the main results of the exploration and analysis of the websites of the government agencies corresponding to the selected countries (concerning the risks of the violation of privacy via videoconferences). In the second part, the main results of the analysis and documentary review are presented.

i. Prevention and reaction of public institutions to the risks of violation of the privacy of users in videoconferences

At the beginning of the global confinement caused by the COVID-19 pandemic, the population tried with the resources at hand, to give continuity to their educational and professional activities through digital media. In this context, Zoom became one of the most used tools. However, there were simultaneously a series of complaints against Zoom for relevant failures and breaches of the security of its users. This fact reflected how vulnerable and unprotected citizens can become in the face of the actions and practices of large technology companies.

Considering the date of the first security alert on Zoom, which caused a worldwide stir, we observe in Table 3.6 that before that date (03/21/2020), none of the web pages analysed of the ministries of education (and other government institutions consulted), offered any guidance to users on how to mitigate the risks of privacy violation when using videoconference (or any related alerts on the (mis)use of the data generated during the video calls).

As stated by Pangrazio and Selwyn (2020), digital technologies have a much faster rate of development than society's ability to analyse their consequences. The governments, therefore, tend to play a role more reactive than proactive on these types of risks to which citizens are exposed. This dynamic can be seen in Table 3.7. When the first alarms go off, six out of eight countries issued security breach alerts to videoconference users.

At the same time, seven countries published documents with recommendations for the safe use of videoconferencing for their population (Table 3.8). Of these seven, only in two countries (Argentina and Ecuador) was the Ministry of Education

TABLE 3.6 Recommendations about user privacy and safe and responsible use, before 03/21/20

	Group 1			*Group 2*			*Group 3*	
	Argentina	*Chile*	*Uruguay*	*Brazil*	*Colombia*	*Peru*	*Ecuador*	*Paraguay*
Education Ministries	No	No	No	No	No	No	No	No
Other government institutions	No	No	No	No	No	No	No	No

TABLE 3.7 Security alerts about the use of videoconferences in education. 03/21/20–04/31/20

	Group 1			*Group 2*			*Group 3*	
	Argentina	*Chile*	*Uruguay*	*Brazil*	*Colombia*	*Peru*	*Ecuador*	*Paraguay*
Government institutions	No	Yes	Yes	Yes	Yes	Yes	No	Yes

TABLE 3.8 Recommendations about risks of violation of the privacy of users in videoconferences. 03/21/20–04/31/20

	Group 1			*Group 2*			*Group 3*	
	Argentina	*Chile*	*Uruguay*	*Brazil*	*Colombia*	*Peru*	*Ecuador*	*Paraguay*
Education Ministries	Yes	No	No	No	No	No	Yes	No
Other government institutions	Yes	Yes	Yes	No	Yes	Yes	Yes	Yes

the entity responsible for issuing this information. In the case of Colombia, Chile, Paraguay, Peru and Uruguay, the "Recommendations" were included in the same "Alerts" document.

As Table 3.8 indicates, in all the countries from Group 1 documentation about safety was found. For instance, explicit recommendations for the safe use of videoconferences.

Over the year, many countries' public institutions generated more security protocols to address these types of threats. However, beyond the level of human development and the capacity of access to digital technologies that countries have (Groups 1, 2 and 3), worryingly limited actions were taken to warn and protect the population against the risks of data violations when using digital services (Raffaghelli, 2020). It is not a coincidence that the digital protection of citizens can be seriously undermined when government institutions don't act effectively. An abundance of literature (Morozov, 2018; Reidenberg and Schaub, 2018; Williamson, 2018) highlights how large technologies penetrate society and particularly education, without adequate technical control and supervision from governments.

Below are some of the main security incidents that Zoom faced during the period studied. Emphasis is also placed on the reaction of public institutions to these threats.

ii. Security incidents: Alerts, locations and recommendations for use

The eight documents selected for this dimension (Table 3.8) are chosen because, while they share relatively similar content, they provide relevant background and concise and straightforward content supported by relevant laws, local regulations, evidence and references.

Most of these documents were published between the last week of March and the first half of April 2020. Although all documents incorporated an overview of the problem, five of these included both a safety alert and recommendations for safe use or data protection guidance (Table 3.8). The cases of the MJPS-Brazil and MINCIT-Colombia documents is relevant (both in Group 2), since at the same time as raising the corresponding alerts and recommendations, they request Zoom Video Communications Inc., to answer a set of questions about the treatment and use of information from users in their respective countries[3].

The reaction of these public organisations reflects an institutional will to respond to a phenomenon that exceeded the anticipation capacity of public and educational entities (Williamson, 2018). Although this reveals a substantive deficiency in the ability to face the underlying risks and violations of privacy, at the same time these institutions showed signs of a deeper understanding of data protection and cybersecurity (Rivera-Vargas and Cobo, 2020).

The analysis identified nine security incidents regarding the use of Zoom. Here we have selected and analysed a subset with the three more frequent (Table 3.9).

iii. User information and credential theft

Reviewing all of the documents, we observed that one of the most common security problems is the vulnerability of Zoom to attacks from various cybercrimes and the vulnerability of user information. This has led to the fact that in different international contexts, Zoom's privacy policy has been permanently questioned (Wagenseil, 2020).

Zoom refers to three aspects considered relevant for this analysis:

- It recognises the collection of a large amount of private information about users: such as name, address, email address, telephone number, position and even employer. It acknowledges that the app collects and maintains data about the type of device one uses and one's IP address, even if the app is used, without creating a Zoom account.
- Zoom certifies that it does not sell the personal data of its users to third parties, but that it does share them to benefit the "business" of those companies.
- Zoom states that it will only be "Responsible for the Processing of Personal Data" related to the provision of the service, but not for the data arising from the interactions during any session that a Zoom user attends, such as recordings, meeting or call transcripts (Zoom, 2020).

Beyond this privacy policy, complaints and reports against Zoom have proliferated since the beginning of the pandemic. The theft of Zoom user credentials offered for sale on the dark web has been one of the most recurrent complaints. In our analysis,

TABLE 3.9 Main identified security incidents

Security incidents	*Documents analysed*								
	Group 1			*Group 2*			*Group 3*		
	EDUCAR-Argentina	*CSIRT-Chile*	*CERT-Uruguay*	*MJPS-Brazil*	*MINCIT-Colombia*	*SGD-Perú*	*TIS-Ecuador*	*MITIC-Paraguay*	*N*
User information and credential theft[1]	X	X	X	X	X	X	X	X	8
Transfer of information to Facebook[2]		X	X		X	X		X	5
Zoombombing[3]	X	X	X	X	X	X	X	X	8
Accelerated growth in the number of users[4]		X		X	X				3
Violation of local laws				X	X	X			3
Zoom Doom[5]	X		X						2
End-to-end encryption in video conferences[6]		X	X					X	3
Zero Days[7]			X		X				2
Counterfeiting and scams[8]			X		X				2

we observed that all the documents refer to this fact. For example, from Group 1, the CSIRT-Chile proposes the following:

> One of the latest relevant vulnerabilities for the Zoom app raises the possibility that third parties can steal the 'login' credentials (or access to your computer) and arbitrarily execute commands on your system. This vulnerability is classified as "UNC path injection".
>
> (p. 8)

From Group 3, the MITIC-Paraguay document mentions the complaint by the Washington Post[4] concerning the discovery of a public website where thousands of videos recorded by Zoom were stored. According to this document: "Many of these videos include personal information, intimate conversations, therapy sessions, physical training, employee training, classes with information for minors and meetings involving financial and confidential information" (p. 1). The document explains that many of these videos were recorded through Zoom and saved in the cloud without a password. Since Zoom names each video recording identically, a simple internet search can reveal a long list of videos that anyone can download and watch[5].

From Group 2, the MINCIT-Colombia document contains a formal request from the Directorate of Personal Data Protection Investigation of this country to Zoom[6], requesting information in this regard. In the same document, the company responded the following:

> [...] no user credentials (e.g., usernames and passwords) were extracted from Zoom. In this regard, any alleged Zoom access credential would actually be a credential used by Zoom users on other websites or applications that were extracted through some security incidents that involved personal data but did not involve Zoom. Some malicious actors now claim that these previously stolen sets of credentials can be used to access Zoom accounts.
>
> (p. 3)

Despite Zoom's response, the documentation from five countries from different groups (Group 1: CSIRT-Chile, CERT-Uruguay; Group 2: MJPS-Brazil, MINCIT-Colombia; Group 3: TIS-Ecuador) raised reasonable doubts about the accuracy of these explanations given by the company. In fact, these five refer to the article published on the Bleeping Computer website[7] which states: "More than 500,000 Zoom accounts are sold on the dark web and forums of hackers for less than a penny each and, in some cases, they are given away for free".

In the MINCIT-Colombia document, the Zoom company stated:

> A small number of valid credentials (significantly fewer than the publicly reported numbers) have been identified as appearing on the dark Internet, and in each case, Zoom has locked the account and required the relevant user to change their password. Zoom generally recommends that its users use

> complex and secure passwords that are not used anywhere else and that have not been exposed in any previous personal data breaches.
>
> (p. 3)

Along with this violation, the documents of Group 1 of countries (CSIRT-Chile, CERT-Uruguay) and MINCIT-Colombia from Group 2, place special weight on Windows credentials. All three suggest that the theft of these would be a vulnerability that attackers could easily perform from the chat of the desktop application. Zoom responded that the company was not aware of any theft of Microsoft credentials that has occurred or affected Zoom's systems (MINCIT-Colombia, 2020). However, in the CSIRT-Chile document it is stated that the Windows Zoom client is vulnerable to UNC[8] path injection in the chat function. This bug could allow attackers to steal the Windows credentials of users who click the link. "All versions for Windows up to 4.6.8 (19178.0323) are vulnerable" (p. 7). In fact, in the article published by the web portal "The Hacker News"[9], all an attacker has to do is to send a created URL to a victim through a chat interface.

It is suggested in the CERT-Uruguay document (Group 1) that to fix this issue, Zoom should prevent the chat system from converting UNC paths into clickable hyperlinks. In this same document, and also in the documentation of the "Group 3" countries (TIS-Ecuador and MITIC-Paraguay), some relevant suggestions are made to enhance the security of the personal data of Zoom users.

iv. Transfer of information to Facebook

Another of the security alerts with the high media relevance linked Zoom with the largest social networks in the world. In this case, the alerts made by countries belonging to Group 2 are relevant. The MJPC-Brazil document refers to an article entitled "Zoom iOS app Sends Data to Facebook Even if You Don't Have a Facebook Account" published on March 28, 2020 on the "Motherboard" website. This article stated that the Zoom application, in its version for iOS, shared its users' data with the social network Facebook, even if the user did not have an account on this network.

Some technical elements of this complaint are provided in the SGD-Peru alert document:

> When the user opens the Zoom application and connects to the Facebook Graph4 API[10], they notify some details of the user's device, among these: the model, time zone, connection territory, telephone operator used, and an identifier of an advertiser created by the user's device.
>
> (p. 10)

Zoom's policy states that the company can collect the user's Facebook profile information (when using Facebook to announce a session in our products or to create an account for our products) but does not explicitly mention anything about the sending of data to Facebook of Zoom users who do not have an account on the social network. However, the MINCIT-Colombia document denounces Facebook's Software Development Kits (SDKs) which were collected data from devices. Like many technology companies, Zoom uses SDKs to add functionality to its applications. An SDK is a set of computer codes that a company provides so that other developers can easily integrate different features provided by that company into the developers' software products (Valdellon, 2020). SDKs typically collect and share information – primarily related to the specifications of the user's device but not personal information.

According to the information published in the MINCIT-Colombia document, the company affirmed the following about the use of SDK to collect user data:

> On March 25, 2020, Zoom learned that SDK was also sharing certain technical information to Facebook, including, operating system type and version, device time zone, device operating system, device model, telecommunication service provider, or screen size. Zoom immediately proceeded to fix this issue and on March 27, 2020, Zoom removed the SDK from the latest version of the Zoom apps. Zoom also asked Facebook to delete any information it had received through the SDK.
>
> (p. 3)

On March 27, 2020, the company published an article on its blog entitled "Using the Facebook SDK Zoom in the iOS client". In it, they claim to have corrected the SDK for the login. As stated in the SGD-Peru document, "Zoom (like other applications) can provide information to third parties to receive information from their application and thus use that information to provide it to measurement services and targeted advertisements" (p. 10).

The information provided by the documentation of countries belonging to Groups 1 and 3 does not provide more background or relevant information than that already provided by countries of Group 2.

v. Zoombombing

As Zoom increased in popularity during the pandemic, there were multiple disruptions to video conferencing. From Group 1, the SGD-Peru document refers to a complaint made by the Boston FBI on March 30, where they warned about the hacking of video conferences in the Zoom app[11]. This phenomenon was known as "Zoombombing". Based on the different emphasis that each document places on the phenomenon of Zoombombing, and especially in relation to the countries of Group 1, we can define Zoombombing as a form of cyberbullying reported by some Zoom users, based on the interception of some calls from unidentified people and trolls, who used hateful language and sometimes shared decontextualised and disturbing images.

Practically all the documents analysed recognised that abusive disruptions in Zoom conferences are not the result of cyber intrusions or illegal accesses. In the case of Group 1 countries, according to the CERT-Uruguay document, "said disruptions are the product of a bad configuration of the session on the part of users, who have publicly shared the identification of the conferences, and in some cases, their passwords if they have been established" (p. 3). The EDUCAR-Argentina document indicated that "the creation of the 'Waiting Room' option in calls from the company was a solution that largely helped mitigate this risk" (p. 1). The Zoom company declared in its blog that it attributes the phenomenon to the substantial increase in non-professional or business users who began to use it once the pandemic started[12]. As stated in the document TIS-Ecuador (Group 3):

> For many users it is their first experience using these available platforms, so the importance of privacy and security policies must be understood, along with the terms and conditions that are accepted when connecting to said platforms.
>
> (p. 1)

Finally, in all the documents analysed, we found substantive recommendations for users to avoid this type of disruption in videoconferences. The document of EDUCAR-Argentina[13] and that of MITIC-Paraguay[14] propose a substantive set of measures to avoid potential incidents.

Discussion and conclusions

The discussion is structured by three questions:

1. ***How did public institutions in South America react when the privacy risks of Zoom were made known?***
   This study highlights how complex the adoption of digital technologies is for education. At the same time, it illustrates some of the challenges that public institutions face when they want to "keep up" with the rapid transformations in the digital landscape. The case of Zoom in Latin American countries illustrates how difficult it can be for public institutions to react in a timely and effective way. Not only do they need to encourage the use of these technologies, but they also need to provide guidance and protect users when they incorporate these tools in their daily life. As Downes and Chunka indicated 20 years ago (1998), in their "law of disruption", technology develops exponentially while social, economic and legal systems change incrementally.

   On a positive note, for the last decade or so the risks associated with privacy have been matters of interest mainly for elite groups: academics, regulators, and other expert communities (e.g., specialised media, civil society advocates, etc.), but have not been matters of concern for non-expert users. To some extent, this is (possibly) starting to change. Given the explosive adoption of social media, the different scandals derived from privacy breaches as well as the growing power of the Big Tech in our society, there is an emergent interest from the broader community to protect themselves from some associated risks. The fact that ministries of education and the governments, in a large number of developing countries (in South America if not beyond) have addressed this matter, opens opportunities to increasing pressure (lobbying) to these companies to change their privacy policies, and also to better educate the citizens in subjects related to privacy and data awareness. However, publishing guidance notes and manuals with recommendations from public institutions will not be enough to mitigate these challenges. The existing asymmetries of power are likely to remain intact if additional measures are not taken (Lanier and Euchner, 2019).

2. ***Where did the public institutions of South America put their emphasis?***
Although the experiences and publications analysed in this chapter are fairly recent, the risks of privacy are not new and certainly did not start with the COVID-19 pandemic. The global lockdown has been followed by an explosive expansion of digital technologies which have increased the tension between the necessity of effectively adopting these digital tools and the importance of minimising unintended consequences (e.g., lack of privacy, surveillance, the concentration of power, manipulation, among others) (Cobo, 2019; Zuboff, 2019).
Our study shows that beyond concrete differences between countries, both Group 1 (Argentina, Chile and Uruguay) and Group 2 (Brazil, Colombia and Peru) reacted with greater promptness to face safety and privacy concerns caused by the massive use of videoconferences during the confinement. Also, both groups generated documentation and guidance assisting citizens to understand the privacy problems. On the other hand, and considering their frontier social development conditions and technological access possibilities, it is observed that citizens from the countries included in Group 3 (Ecuador and Paraguay) were in a situation of greater vulnerability, and probably of a more constrained "technological readiness" (Kirikkaleli, et al. 2021) exemplified by limited guidance and protection in terms of how governments reacted to prevent digital threats.

3. ***What lessons can be learned?***
Protecting citizens in general (and students in particular) from the risks associated with the use of Zoom by simply providing quick privacy tips will not be enough. The lessons from 2020 show that it is important to have a deeper understanding of what it means to be online.
Being connected is a practice of deep relevance and essential part of contemporary life. Using Zoom (or similar) has become a necessity for many students (when access to the internet is available). On the other hand, being offline is not even an option for most users particularly during the lockdown (what used to be a privilege a few decades ago is now a basic need in most modern societies). For instance, those users who did not have access to digital technologies have paid a very high cost during the pandemic (e.g., isolation, exclusion, disconnection from their schools and peers, etc.). The centrality of digital tools in today's society makes it difficult to adopt a critical position. However, even when acknowledging the benefits of digital technologies there is still space to

highlight the social and political challenges posed by these tools. Ignoring or simply minimising these challenges could amplify existing inequalities.

A growing number of authors are emphasising the importance of consolidating institutional and individual capacities to learn how to navigate (more) safely the current landscape. Some of these voices reinforce the importance of citizens' deliberation and the importance of being proactive, for instance to adopt more severe sanctions to those who undermine people's privacy, to audit how public and private providers use citizens' data, to promote the creation of ombudsman services, etc. Having strong digital governance is essential. A form of governance that is capable not only of reacting to threats that challenge the safety of citizens, but that can anticipate the risks and educate conscientious and critical citizens in the digital society. Our study shows that none of the three groups of countries analysed were sufficiently prepared to face these threats to the privacy of people who actively used videoconferencing platforms to continue their training and professional processes. The effects from the loss of privacy in digital society is still an open question (e.g., erosion of freedom, potential abuses derived from misusing individuals' personal data, unscrupulous use of information[15], manipulation of people's behaviour, among others) (Eyal, 2014). Together this highlights the importance of educating citizens as (pro) active, resilient, smartly sceptical (Marantz, 2020), and critical thinkers (Buckingham, 2019).

The study of Zoom in South American countries is one example which underscores the necessity to revisit regulations, transform existing digital governance (Berman and Albright, 2017), and in many cases to increase awareness (or education) of citizens (Livingstone, 2018). In addition to more protective regulations, this study emphasises the necessity to develop a more advanced set of (critical) digital capacities (Raffaghelli, 2020) to better prepare individuals not only to understand these challenges but to also take pre-emptive actions (Hartung, 2020). If we only look at the positive effects of a digital society but ignore its unintended consequences, we are inaugurating a new form of digital ignorance.

Finally, it is quite evident that the education sector strongly benefits from incorporating digital services offered by these global tech providers. But it is important not to overlook the role education institutions have adopted (consciously or not) as major data providers. Vast amounts of personal data are collected every day from schools and education users to train algorithms that mostly increase the profit margins of large tech companies. As noted by Selwyn (2016) school systems have relatively unsophisticated data processing technologies but also limited data governance plans that are, in most cases, ill-prepared to face these challenges. The enthusiasm generated with the new technologies in education has not gone hand in hand with the consolidation of a robust digital governance capable of protecting citizens from the threats to privacy, data manipulation or other derivative risks. In addition to the adoption of new tools and sophisticated algorithms, education systems need to build time-proof institutional capacities, regulations and guidance to protect themselves from the unintended consequences of the digital revolution.

Notes

1 In the case of Brazilian institutions, the search was carried out in Portuguese. In the case of the rest of the countries included, the search was done in Spanish.
2 See details here: https://www.google.com/advanced_search?q=google&sxsrf=ALeKk01ztZdqDVJ2QaLKraI6xR8rJnBZnQ:1607652617128&hl=es
3 In the case of the MINCIT-Colombia document, this also includes the responses of Zoom Video Communications Inc. to the subpoenas raised by the MINCIT.
4 Retrieved from: https://www.washingtonpost.com/technology/2020/04/03/thousands-zoom-video-calls-left-exposed-open-web/
5 This problem occurs because some users decide to record their meetings and choose the option to "record and save" in the cloud. It should be noted that Zoom does not make these recordings by default, and users who have chosen to "record and save" their meetings on their computers are not affected by this privacy issue.
6 A summons made on April 14, 2020 by the Personal Data Protection Investigation Directorate of MINCIT-Colombia is reproduced, requesting information in this regard.
7 See article here: https://www.bleepingcomputer.com/news/security/over-500-000-zoom-accounts-sold-on-hacker-forums-the-dark-web/
8 Short for Universal Naming Convention or Uniform Naming Convention, it is a PC format for specifying the location of resources on a local-area network (LAN). UNC uses the following format: \\server-name\shared-resource-pathname. So, for example, to access the file test.txt in the directory EXAMPLES on the shared server SILO, you would write: \\SILO\EXAMPLES\test.txt. See: https://kb.blackbaud.com/articles/Knowledge/48745
9 The aforementioned article includes the tests carried out by "Mohamed Baset" in which you can see how the attack works. https://thehackernews.com/2020/04/zoom-windows-password.html
10 API GRAPH is the main way developers get data on or off Facebook.
11 See alert here: https://www.fbi.gov/contact-us/field-offices/boston/news/press-releases/fbi-warns-of-teleconferencing-and-online-classroom-hijacking-during-covid-19-pandemic
12 See details here: https://blog.zoom.us/keep-uninvited-guests-out-of-your-zoom-event/
13 See details here: https://www.educ.ar/recursos/154545/cuidados-al-momento-de-armar-una-clase-virtual-con-videoconf
14 See details here: https://www.mitic.gov.py/noticias/multiples-problemas-de-privacidad-en-zoom
15 See documentary: "The Social Dilemma". Directed by Jeff Orlowski. Exposure Labs, Argent Pictures and The Space Program, 2020. *Netflix*, https://www.netflix.com/uy/title/81254224

References

Ahlquist, J. (2016). The digital identity of student affairs professionals. *New Directions for Student Services*, *155*, 29–46. doi: 10.1002/ss.20181

BBC News Mundo (2020, March 25) Coronavirus: Por qué Zoom se ha vuelto tan popular para realizar videollamadas (y otras 3 aplicaciones gratuitas). *BBC*. Retrieved from: https://www.bbc.com/mundo/noticias-52040148

Berman, G. & Albright, K. (2017). *Children and the Data Cycle: Rights and Ethics in a Big Data World. UNICEF Office of Research-Innocenti*. Retrieved from: https://www.unicef-irc.org/publications/pdf/IWP_2017_05.pdf

Bernal, P. (2020). *What Do We Know and What Should We Do About Internet Privacy?* London: SAGE.

Brewczyńska, M. (2020). Poland. The Polish government's actions to fight COVID-19: A critical look at the 'selfie app' and direct access to location data. *European Data Protection Law Review*, *6*(2), 301–307. doi: 10.21552/edpl/2020/2/17

Buckingham, D. (2019). *The media education manifesto*. New Jersey: John Wiley & Sons.

Burch, P., & Miglani, N. (2018). Technocentrism and social fields in the Indian EdTech movement: Formation, reproduction and resistance. *Journal of Education Policy*, *33*(5), 590–616. doi: 10.1080/02680939.2018.1435909

Calzada, I., & Cobo, C. (2015). Unplugging: Deconstructing the smart city. *Journal of Urban Technology 22*(1), 23–43. doi: 10.1080/10630732.2014.971535

Cho, J., Saifuddin Ahmed, S., Hilbert, M., Liu, B., & Luu, J. (2020) Do search algorithms endanger democracy? An experimental investigation of algorithm effects on political polarization. *Journal of Broadcasting & Electronic Media*, *64*(2), 150–172, doi: 10.1080/08838151.2020.1757365

Choe, Y. J., & Choi, E. H. (2020). Are we ready for coronavirus disease 2019 arriving at schools? *Journal of Korean Medical Science*, *35*(11), 1–2, doi: ggq848

Cobo, C. (2019). *Acepto las condiciones: Usos y abusos de las tecnologías digitales*. Fundación Santillana.

Coursera (2020, March 25) Coursera together: Free online learning during COVID-19 [Blog] *Coursera Blog*. Retrieved from: https://blog.coursera.org/coursera-together-free-online-learning-during-covid-19/

Delgado, P., Illanes, R., & Lucio-Villegas, E. (2020). Políticas educativas en tiempo de coronavirus. *Laplage em Revista*, *6*(2), 4–6. doi: 10.24115/S2446-6220202062857.

Digital Future Society (2019). Privacy-first: A new business model for the digital era. Retrieved from: https://digitalfuturesociety.com/app/uploads/2020/03/THINK_TANK_I6_Privacy-first_a-new-business-model-for-the-digital-era_ENG.pdf

Downes, L., & Mui, C. (1998). *Unleashing the killer app: Digital strategies for market dominance*. Cambridge, MA: Harvard Business School Press.

Edwards, D., & Potter, J. (1992). *Discursive psychology*. London: Sage.

Emejulu, A., & McGregor, C. (2019). Towards a radical digital citizenship in digital education. *Critical Studies in Education*, *60*(1), 131–147. doi: 10.1080/17508487.2016.1234494

Encinas-Martín, M. (2020), *Spain: Aprendo en casa (Learn at home), Education continuity stories series*, OECD Publishing. Retrieved from: https://bit.ly/2YjUNHP

Espinoza, A. (2020, March 19) Microsoft dará acceso gratis a Microsoft Teams durante pandemia de COVID-19. *It Now*. Retrieved from: https://revistaitnow.com/conomía-dara-acceso-gratis-a-microsoft-teams-durante-pandemia-de-covid-19/

Eyal, N. (2014). *Hooked: How to build habit-forming products*. New York: Penguin Random House.

Falzon, M. A. (2009). *Multi-sited Ethnography: Theory, praxis and locality in contemporary research*. Ashgate.

Fardella, C., Sisto, V., & Jiménez, F. (2017). La transformación de la universidad y los dispositivos de cuantificación. *Estudos de Psicologia (Campinas)*, *34*(3), 435–448.

Federal Trade Commission (2020, November 9) *FTC Requires Zoom to Enhance its Security Practices as Part of Settlement [Press release]* Retrieved from: https://www.ftc.gov/news-events/press-releases/2020/11/ftc-requires-zoom-enhance-its-security-practices-part-settlement

Fernández, E. (2019, December 21) La tasa Google: Todo el mundo busca 90.000 millones en impuestos de los gigantes de Internet. *El Mundo*. Retrieved frome: https://www.elmundo.es/economia/empresas/2019/12/21/5dfd24cdfdddff84108b467a.html

Gallegos Cárdenas, M.A. (2020 May 22) ¿Por qué Google for Education es útil?. *La Vanguardia Daily*. Retrieved from: https://www.lavanguardia.com/participacion/lectores-corresponsales/20200522/481302863100/suite-google-for-education-implantacion-escuelas-ciudad-de-mexico-pandemia-covid-19.html

García, J.G. (2020, April 07) Los problemas de privacidad y seguridad sacuden el éxito de Zoom. *El Pais: America*. Retrieved from: https://elpais.com/retina/2020/04/06/innovacion/1586169223_190951.html

García, S. (2017). Google y su abuso de posición. *RUE: Revista universitaria europea*, *27*, 143–162. Retrieved from: http://www.revistarue.eu/RUE/122017.pdf

Hartung, P. (2020) The children's rights-by-design standard for data use by tech companies. Good Governance of Children's Data project Office of Global Insight and Policy. Issue brief, 5: *Unesco*. Retrieved from: https://www.unicef.org/globalinsight/media/1286/file/%20UNICEF-Global-Insight-DataGov-data-use-brief-2020.pdf

Hine, C. (2004). *Etnografía virtual*. UOC.

HolonIQ (2020, August 6). Global EdTech Market to reach $404B by 2025 – 16.3% CAGR. Retrieved from: https://www.holoniq.com/notes/global-education-technology-market-to-reach-404b-by-2025

Hutchinson, A. (2020, May 28) TikTok Announces #LearnOnTikTok Initiative to Encourage Education During Lockdowns. *Social Media Today*. Retrieved from: https://www.socialmediatoday.com/news/tiktok-announces-learnontiktok-initiative-to-encourage-education-during-lo/578805/

Ibáñez, T., & Iñiguez, L. (1996). Aspectos metodológicos de la psicología social aplicada. In J. Álvaro & A. Garrido (Eds.), *Psicología social aplicada* (pp. 57–82). McGraw-Hill.

Iñiguez, L., & Antaki, C. (1994). El análisis del discurso en Psicología. *Boletín de Psicología*, 44, 55–75.

ITU (2018). *Measuring the information society report*. Geneve: ITU Publications.

Jarke, J, & Breiter, A. (2019). The datafication of education. *Learning, Media and Technology*, *44*(1), 1–6.

Kari, P. (2020, June 17) Zoom will provide end-to-end encryption to all users after privacy backlash. *The Guardian: US*. Retrieved from: https://www.theguardian.com/technology/2020/jun/17/zoom-encryption-free-calls

Kirikkaleli, D., Ertugrul, H. M., Sari, A., Ozun, A., & Kiral, H. (2021). Quality of education and technological readiness: Bootstrap panel causality analysis for Northern European countries. *Scandinavian Journal of Educational Research*, *65*(2), 276–287.

Komljenovic, J. (2020). The future of value in digitalised higher education: Why data privacy should not be our biggest concern. *Higher Education*. doi: 10.1007/s10734-020-00639-7

Lanier, J., & Euchner, J. (2019). What has gone wrong with the internet, and how we can fix it: An interview with Jaron Lanier. *Research-Technology Management* *62*(2). 13–20. doi: 10.1080/08956308.2019.1587284

Livingstone, S. (2018). Children: A special case for privacy? *Intermedia*, *46*(2). 18–23. Retrieved from: http://eprints.lse.ac.uk/89706/

Mack, Z. (2019, April 9). Jaron Lanier's ideas for the future of profiting from your own data. *The Verge*. Retrieved from: https://www.theverge.com/2019/4/9/18302076/data-monetization-control-manipulation-economy-jaron-laniers-virtual-reality-vr-vergecast

Marantz, A. (2020). *Antisocial: Online extremists, techno-utopians, and the hijacking of the American conversation*. Penguin Random House.

Menéndez, J. (2020, June 24) Exfiltración de datos mediante Google Analytics. *Una al Día*. Retrieved from: https://unaaldia.hispasec.com/2020/06/exfiltracion-de-datos-mediante-google-analytics.html

Miño-Puigcercós, R., Rivera-Vargas, P., & Alonso Cano, C. (2019). Comunidades virtuales: dinámicas emergentes de participación social y aprendizaje entre los jóvenes. *Education in the Knowledge Society*, *20*, 1–12. doi: 10.14201/eks2019_20_a21

Morozov, E. (2018). *Capitalismo Big Tech: ¿Welfare o neofeudalismo digital?* Madrid: Enclave.

Morrison, S. (2020, December 23) The year we gave up on privacy. *Vox Media*. Retrieved from: https://www.vox.com/recode/22189727/2020-pandemic-ruined-digital-privacy

O'Flaherty, K. (2020, October 14) Zoom beats microsoft teams, google meet with game-changing new features. *Forbes*. Retrieved from: https://www.forbes.com/sites/kateoflahertyuk/2020/10/14/zoom-beats-microsoft-teams-google-meet-with-game-changing-new-features/?sh=673463ad6cba

Pangrazio, L. & Selwyn, N. (2020) Towards a school-based critical data education. *Pedagogy, Culture & Society*, 1–18. doi: 10.1080/14681366.2020.1747527

Pardo, M., Waliño-Guerrero, M., & San Martín, Á. (2018). La "uberización" de los centros escolares: Reorganización del trabajo pedagógico mediante las plataformas digitales de contenidos. *Educatio Siglo XXI*, *36*(2), 187–208. doi: 10.6018/j/333031

Potter, J. (1998). *La representación social de la realidad. Discurso, retórica y construcción social*. Paidós.

Prior, L. (2008). Repositioning documents in social research. *Sociology*, *42*(5), 821–836. doi: 10.1177/0038038508094564

Raffaghelli, J. (2020). Is data literacy a catalyst of social justice? A response from nine data literacy initiatives in higher education. *Education Sciences*, *10*(9), 233. doi: 10.3390/educsci10090233

Regan, P.M. & Jesse, J. (2019). Ethical challenges of EdTech, big data and personalized learning: twenty-first century student sorting and tracking. *Ethics and Information Technology*, *21*, 167–179. doi: 10.1007/s10676-018-9492-2

Reidenberg, J. & Schaub, F. (2018). Achieving big data privacy in education. *Theory and Research in Education* *16*(3), 263–279. https://journals.sagepub.com/doi/abs/10.1177/1477878518805308

Rivera-Vargas, P., & Cobo-Romani, C. (2020). Digital learning: Distraction or default for the future. *Digital Education Review*, (*37*), 1–16. doi: 10.1344/der.2020.37.%25p

Sádaba, I. (2016). Capitalismo cognitivo y Sociedad de la Información: De la innovación la Bid Data, *Con-Ciencia Social*, *20*, 21–30.

Sancho-Gil, J. M., Rivera-Vargas, P., & Miño-Puigcercós, R. (2020). Moving beyond the predictable failure of EdTech initiatives. *Learning, Media & Technology*, *45*(1), 61–75. doi: 10.1080/17439884.2019.1666873

Schwartzman, R. (2013). Consequences of commodifying education. *Academic Exchange Quarterly*, *17*(3), 41–46.

Secara, I. A. (2020). Zoombombing–the end-to-end fallacy. *Network Security*, *2020*(8), 13–17. doi: 10.1016/S1353-4858(20)30094-5

Selwyn, N. (2016). 'There's so much data': Exploring the realities of data-based school governance. *European Educational Research Journal*, *15*(1), 54–68.

Snowden, E. (2019). *Permanent record*. Macmillan.

UNESCO (2020). COVID-19: 10 Recommendations to plan distance learning solutions. *UNESCO*. Retrieved from: https://cutt.ly/dyNYGPw

Tracy, R. (2020, December 17) More states hit google over alleged monopoly conduct. *The Wall Street Journal: American Edition*. Retrieved from: https://www.wsj.com/articles/states-file-antitrust-lawsuit-against-google-11608228334

Valdellon, L. (2020, July 30). What is an SDK? Everything you need to know. *ClaverTap*. Retrieved from: https://clevertap.com/blog/what-is-an-sdk/

van Veen, E. (2018). Observational health research in Europe: Understanding the General Data Protection Regulation and underlying debate. *European Journal of Cancer, 104*, 70–80. doi: 10.1016/j.ejca.2018.09.032

Véliz, C. (2020) *Privacy is Power: Why and How You Should Take Back Control of Your Data*. Penguin Random House.

Wagenseil, P. (2020). Zoom security issues: Here's everything that's gone wrong (so far). *Tom's Guide, 11*. Retrieved from: https://www.tomsguide.com/news/zoom-security-privacy-woes

Wetherell, M., & Potter, J. (1998). El análisis del discurso y la identificación de los repertorios interpretativos. In A. Gordo & J. Linaza (Eds.), *Psicología, discurso y poder: Metodologías cualitativas, perspectivas críticas*. Visor.

Williamson, B. (2016). Digital methodologies of education governance: Pearson plc and the remediation of methods. *European Educational Research Journal, 15*(1), 34–53. doi: 10.1177/1474904115612485

Williamson, B. (2018). *Startup Schools, Fast Policies, and Full-Stack Education Companies. The Wiley Handbook of Global Educational Reform*, 283–305. doi: 10.1002/9781119082316.ch14

Williamson, B., & Piattoeva, N. (2019). Objectivity as standardization in data-scientific education policy, technology and governance. *Learning, Media and Technology, 44*(1), 64–76. doi: 10.1080/17439884.2018.1556215

World Bank (2019). *Individuals using the Internet (% of population), global report*. Washington, DC: World Bank data.

Zoom Privacy Statement (2020, August). Retrieved from: https://zoom.us/privacy

Zuboff, S. (2019). Surveillance capitalism and the challenge of collective action. *New Labor Forum, 28*(1), 10–29. doi: 10.1177/1095796018819461

4

DATA CLASSES

An investigation of the people that 'do data' in schools

Neil Selwyn, Luci Pangrazio and Bronwyn Cumbo

Introduction

Schools are now sites for the voluminous production of digital data – from external examination and assessment grades, through to internal measures of student attendance, parental satisfaction and teacher performance. While schools have always collected and recorded non-digital data on students through examinations and attendance registers, the rise of powerful data processing systems has fuelled expectations for schools (and the people who work in schools) to be "data-driven" and "data-intensive" in everything that they do. All told, schools and classrooms are being recast as environments where "it is no longer acceptable to simply use anecdotes, gut feelings, or opinions as the basis for decisions" (Mandinach, 2012, p.71). Instead, school staff (at all levels of seniority) are expected to engage in "data-driven decision-making" as a means of improving their practice and ensuring their accountability.

This "data turn" is generally portrayed in positive terms in educational spheres. While it is acknowledged that schools' actual data use remains inconsistent, engagement with data nevertheless continues to be promoted as a basis for the improvement of the school system. School leaders are now encouraged to foster institutional "data cultures" (Schildkamp et al., 2017) and to establish shared norms for data use. Mechanisms for doing this include the development of school-specific "data plans" and the coordination of "data teams" that allow staff to engage with data on collaborative and communal terms (see Wardrip and Herman, 2018; Schildkamp and Gummer 2013). All told, the increasing generation, circulation and process of educational data is associated with promises of significant benefits and reform. As Jarke and Breiter, 2019 put it:

> Digital data allow for the analysis of different educational practices to a degree of complexity not previously possible and to a much greater extent, as they

DOI: 10.4324/9781003136842-4

> can be very detailed, cover a more complete scope and can be flexibly combined [...] digital data not only serve to support decisions, but also fundamentally change the organisation of learning and teaching.
>
> (Jarke and Breiter, 2019)

While the 'grey' literature and policy recommendations for school improvement continues to be driven largely by enthusiastic efforts to support "better" uses of data within schools, an emerging body of work is beginning to examine critically the ways in which this prevailing "logic of enumeration" (Hardy 2015) is actually being played-out *within* individual schools. For example, such studies have illustrated how expectations of sustained data use can intensify teachers' work, lead to diminished professional autonomy and a heightened sense of continuous performativity (see Kim, 2018; Bradbury and Roberts-Holmes 2017; Selwyn et al. 2016).

This chapter seeks to extend these studies by pursing an organisation-focused approach to examining the realities of how school staff are "doing" data. In particular, we take a lead from the emerging "critical data studies" tradition (Kitchin and Lauriault, 2014; Iliadis and Russo, 2016). This encourages us to problematise the rising societal prominence of data (particularly in digitally mediated forms) in terms of associated shifts in power and politics. In this sense, critical data studies draw attention to the fact that the control and use of digital data is becoming a key source of power in society. As such, this raises questions over which individuals and institutions are becoming (more) empowered by their jurisdiction over digital data (and, conversely, who is becoming disempowered). At the same time, it also draws attention to the extent to which data and data-driven processes are becoming "new power brokers" (Diakopolous 2013, p.2) in their own right.

In an educational context, therefore, it makes sense to reassess the rising prominence of school data in these messier critically minded terms. Indeed, Jarke and Breiter (2019) contrast enthusiasms for the inclusive and consistent benefits of school data with the likely divides and inequalities implicit in the competences required to work with and interpret data. If nothing else, differences in technical "know-how" are likely to see the increased involvement of particular staff in schools' data processes, and the subsequent marginalisation of others. As Ian Hardy has previously noted, it makes sense to see professional association with numbers and statistics within a school as a new form of power due to "the seductiveness of the constitution of an enumerate subject position" (Hardy 2015, p.30).

On one hand, then, we might well expect to find a number of newly formed "data classes" within a school. The "data classes" metaphor is multileveled in that it refers not only to how the different groups of teachers understand and work with digital data in schools, but also highlights how these classes lead to new forms of 'data power' (Kennedy and Bates, 2017), which govern indirectly and without proximity. Who gets to "do" data within a school context (and, conversely, who does not) therefore raises new and important questions, particularly with regard to the effect this has on the design of curriculum and pedagogy schools. These might be ordered along lines of technical and statistical expertise – reflecting what Lev Manovich (2011) has described previously as a "data analysis divide" between data experts and those without specialist computational training and skills. At the same time, however, we need to be mindful of the ways in which this remains entwined with broader micropolitics of everyday schooling (see Ball, 1987), such as the school's executive structure, faculty and departmental divisions and other established hierarchies and power dynamics.

Given the focus of this book on datafication *and* education, we are particularly interested in how teachers understand data and how this shapes their professional practices and interactions with students, teachers and administrators. In particular, how data is "done" in schools has an impact on any conceptualisation around pedagogy and curriculum, which of course affects how children and young people learn and understand the world and their place in it.

In this sense, the chapter now goes on to consider the following research questions:

1. Which staff are associated with data use in schools?
2. What motivates these staff to respond to the prioritisation of data within the current educational context?
3. What do these staff see themselves as doing with data (and, at the same time, how do others view their work)?

We begin by outlining the research approach and methods. We then present the findings, which are organised according to the different 'data classes' that we found in schools, namely: executive/senior leadership; data teams; data experts; data enthusiasts; and the 'run-of-the-mill' majority of teachers. The discussion begins by outlining the importance of context, highlighting how the micropolitics of each school shaped the formation of these data classes and the impact on curriculum and pedagogy. Also discussed is how data, which was often conflated with datafication, was treated as an object that could be manipulated and controlled to achieve particular ends. The chapter concludes by speculating on what these new power dynamics mean for teachers and students in Australian schools.

Research approach and methods

With these questions in mind, we now go on to develop a critical account of the digital data work currently taking place within secondary schools in the Australian state of Victoria. To do this, we draw on interview data generated from in-depth qualitative studies of digital data practices in three contrasting Victorian secondary schools across the 2019 academic year. Although we readily acknowledge that schools have always generated and used non-digital data, we do not account for or explore these practices in the present study. This empirical data constituted the opening phase of a three-year research project seeking to make use of participatory methods to reimagine schools' datafied practices. The three case-study schools were a small inner-city government school (Weston High School), a large suburban Catholic school (Brookdale High School), and a medium-sized outer-suburbs private school (Northland College). In Australia, approximately one in five students attend a Catholic school. The Catholic education system is different from both the public and private education systems in that it was established by the Catholic Church. Catholic schools are ostensibly faith communities based on belief in God and a Christian way of life.

This period of immersive fieldwork in the three schools involved identifying and examining the main forms of data generation associated with school ethnography – including over 60 site visits, in situ observations and general "hanging around", field notes, documentary analysis, photographing, corridor conversations and more formal interviews with IT staff, data specialists, school leaders, teachers and students. In particular, this chapter draws upon interviews with over 50 staff across the three schools in teaching, leadership, technology and administrative positions. Each participant is given a pseudonym, which is outlined in the Endnotes to this chapter. These interviews were in-depth and semi-structured, lasting anywhere between 30 and 90 minutes. Drawing on thematic analysis of this corpus, structured initially by the a priori themes of the three broad research questions described above, this chapter now goes on to explore the different forms of data "people" and data "work" that were described by interviewees as constituting how data was being "done" within each school. This can be described in terms of the following five distinct hierarchies of data engagement.

Findings

"Data" as an executive-level concern

The schools' use of data was initially presented to us as a senior leadership concern. Indeed, the influence of Principals, their Deputies and other executive staff was evident across the three schools – each responding to different perceived data imperatives. For example, Weston's Principal Sarah described herself as motivated primarily by the ways in which Victoria's state Department of Education and Training (the "DET") was prioritising data-driven schooling ("things from the Department that are imperatives to do"). This included various directives, reporting criteria and online portals designed to ensure that all government schools were engaging with official data flows. While Sarah was well aware of the importance of data to the DET, she couched this in rather sarcastic terms ("I guess that the basic mantra in education now is 'Understand Thy Impact'"). Through various initiatives and communications from the DET, Sarah was well aware of the fact that data engagement was a principal's responsibility. Nevertheless, Sarah saw 'data engagement' as enabling collective action throughout her staff. For example, she was sharing access to the DET "Principal Data Portal" with a wide range of colleagues beyond the officially permitted five-person limit. As Sarah reasoned:

> I've actually shared the data […] I understand why you wouldn't just give it to everyone [but] I certainly think people in leading roles need to have their own access […] It's [about] teacher agency […] it's not rocket science!
>
> ([1] in Table 4.1)

Brookdale's executive staff, in comparison, described themselves as responding directly to a recent school inspection that had deemed the school to be "failing" in its use of data. While indignant about being criticised by "elderly ex-teachers coming in to tell us how we are doing" they accepted the judgement that the school was "data rich and usage poor" ([2] in Table 4.1). As a result, the school's Headteacher had appointed one of his Deputies as the school's official "Head of Data". Les interpreted this role as requiring him to instigate a "root and branch" reform of the school's IT systems ([3] in Table 4.1). This was viewed by other Brookdale staff with scepticism:

> So, now we have a Dean of Data [who] doesn't know data [...] I'm not having a crack – he's a wonderful man, but that position is a compromise to keep someone in a job who lives and breathes Brookdale [...]. But each year *we* have to show *him* how to upload data [to the LMS].
>
> ([4] in Table 4.1)

In contrast, there was little overt criticism of the ways in which the Executive Principal and his Deputy were leading Northland's ambitions towards becoming a "data-informed" school. From our initial visits onwards, this was presented as a top-down commitment that was shared throughout the school. Indeed, the Principal was confident that "the staff here are open to data" ([5] in Table 4.1). Principal Dennis had been in the post for five years and now felt that "we've got the school to run well as a school normally should". As such, his ambitions to refine Northland's data use were rationalised as "tweaking to get the best out of the machine now the machine's running reasonably well, to use an industrial metaphor" ([5] in Table 4.1). While this was usually talked about in a positive light, Dennis occasionally expressed anxieties over not fully using the mass of data that was being generated within the fee-paying school:

> One of our weaknesses is we've got so much data that we don't use [...] or we just have a cursory look at [...] So I'd hate at some stage to have something go badly wrong and we could've foreseen it.

The formation of "data teams"

Despite their varying levels of assuredness, these "executive" school leaders had little personal expertise regarding data use. For example, Dennis saw himself as contributing experience and "soft skills" to Northland's data agenda ([5] in Table 4.1) – establishing connections and initiating communication amongst school staff. Northland's commitment to improved use of data therefore fell to a formally constituted "Data Leadership Team". This was an eight-person subgroup of the school's "Teaching and Learning Committee". Joining Dennis and his Deputy Principal (Russ) were Northland's IT manager and a second member of the school's "tech team" with responsibility for programming. Three other places fell to senior teaching staff with assigned responsibilities as "project managers" for various aspects of school improvement. The project portfolios deemed to be aligned with data use were the lead staff for "Assessment and Reporting", "Character Assessment", and "Professional Development"). Making up the numbers was Stephen – Northland's "Senior Data Manager". In reality, this group was rarely convened en masse. Rather, various combinations of these staff met when directed. These meetings and interactions were largely initiated by Dennis and Russ, as well as by the school's chief technology officer (CTO). In this sense, the overall "Data Leadership Team" was less of a working team and more the pretence of a working team. Indeed, the "Data Leadership Team" could be read in terms of hierarchy and exclusion rather than indicating any kind of knowledge or expertise.

In contrast, Weston's principal had assembled a similarly sized group of staff that did gather together to "talk about data" ([1] in Table 4.1). Variously referred to as the school's "data group", "data committee" and "data team", this was a more organic arrangement than in Northland. As Sarah reflected, "the first thing we realised when we got together was that everyone has a different take on what 'data' is". This group met together on an informal basis – "we perhaps don't meet as often as we should" ([1] in Table 4.1) – and comprised of six staff with responsibility for teacher professional development, student learning and timetabling.

Reflecting Sarah's ambition to share data beyond the schools' executive, the group's initial interests had settled around the idea of improving data accessibility and data awareness throughout the school. As Vic put it, "looking at how data is used around the school and making it available for other teachers". One point of consensus was the goal of making data a collective responsibility. As per DET protocol, Weston's teachers were arranged in different "professional learning communities" (PLCs) organised around subject specialisms, such as numeracy, literacy and health and physical education. As such, "the idea of the [data] group is to build the capacity of the PLCs – not just the individual teachers – to utilise data" ([6] in Table 4.1). Another tentative aim was to develop way of 'students understanding their own data' and encouraging 'students to begin to take data on board' ([1] in Table 4.1).

On the face of it, Brookdale had a similar group of senior staff with a data remit, albeit with little sense of collective responsibility or cohesion. Some of the group had designated titles. For example, Les was the "Head of Data", while a recently appointed recruit with a background in corporate business intelligence was the "Data Analyst". Other staff involved were the school's "Director of IT", and another senior teacher working in a "Director of Teaching and Learning" role with additional responsibility for the school's learning management system. In contrast to the (semi)formal "team" structures at Northlands and Weston, Brookdale's staff were working along relatively separate lines, with various people coming together to enact Les's reform of the school's data infrastructure. This primarily involved improving interoperability between the school's various data systems. As Les reasoned, "you can't do the data part of the job [...] unless you've got the other parts working" ([3] in Table 4.1).

Officially designated "data experts"

While providing designated responsibility for 'data' within the organisational structure of each school, these different teams and committees offered limited direction and expertise with regard to digital data. For example, we first visited when Weston's team was pursuing the idea of producing "data packs" ([6] in Table 4.1) for each teacher community. These eventually took the form of a series of Google Documents with lists of relevant resources. In all schools, then, any advanced data work (particularly data processing and analysis) was being undertaken by specific individuals who were deemed to have sufficient levels of data expertise, both technically and conceptually.

In Northland and Weston this work fell to two similarly positioned senior teachers. Tony and Stephen were both recognised informally as their schools' "data experts" who took on responsibility for running analyses. This data work was certainly not the result of large analytics systems and complex calculations. Instead, Northland's Senior Data Manager (Stephen) was at the heart of most data analyses within the school, while Weston's "Data Manager" (Tony) fulfilled a similar role. They were both full-time teachers in their mid-fifties teaching mathematics (Stephen) and science (Tony). Both had self-professed interests in data and statistics and had accordingly worked their way into data-focused roles. Tony was Weston's timetabling manager, while Stephen had an allotment of one-day-a-week to improve student "growth" through data (i.e., coordinating the school's efforts around improving student performance). These teachers were entitled to an allotment of non-teaching time to take responsibility for their schools' data analysis.

Both teachers were well-regarded by their colleagues. Northland staff would regularly refer to Stephen as the school's "data guy". Tony was introduced to us by Weston's principal as "very clever with spreadsheets" ([1] in Table 4.1). In contrast, Stephen was quick to describe his skillset in more modest terms: "data and spreadsheeting is something I've just picked up over time [...] I've just sort of fallen into it" ([8] in Table 4.1). Tony was similarly modest: "I don't have any training in it [...] I've just always liked numbers [...] all I do is essentially old-fashioned spreadsheet work in the same way that I was doing ten years ago. It's just me and my laptop" ([9] in Table 4.1).

Both Stephen and Tony had been given responsibility for running the few big data analyses that formed the backbone of data work in Northland and Weston. For example, Stephen ran an annual "prediction exercise" at the beginning of students' final year to produce "robust estimations" of their likely university admission scores. He also ran analyses to identify top-performing students who might be awarded "academic prizes". At Weston, Tony undertook similar analyses, as well as annual reviews of schoolwide data relating to student "lates" and non-attendance. All these data exercises took considerable amounts of time and "manual" effort – meaning that each analysis was conducted only once or twice each year. As such, their data work had become an important part of school knowledge, with other teachers reliant on them to mediate and communicate the analysis of student data. Despite this, both teachers acted largely independently of their colleagues. As Stephen put it, "I keep telling them [...] if I walk under a bus – this is all gone!". Similarly, as one of Tony's colleagues put it, "if only we had two or three of him, it would be amazing" ([10] in Table 4.1).

In contrast, schoolwide data work within Brookdale were designated officially to two staff members. The school's main annual review of examination data fell under Les's purview as "Head of Data". Les took responsibility for uploading and coordinating access to this data. In contrast to Weston and Northland, he did not devolve responsibility for the subsequent analysis to statistically minded colleagues. Instead, this work was outsourced to an external data consultant – a former teacher who teachers described as having a "good reputation" across Melbourne schools. Margaret was currently employed by an independent education research organisation, meaning that Brookdale could officially contract her to run annual analyses of the school's examination data. As Les explained, the value of such a service was bolstered by the opportunity to gain a sense of how Brookdale compared with the research organisation's other clients:

> One good thing about Margaret is that she also tells you how you rank against other schools as well. She does this for about 40 or 50 schools. So we can see how our classes did against them [...] Of course, the problem with that is that Margaret charges $2,000 to do this! But we consider it pretty important, and we've done it for a long, long time. There'd be a lot of schools who would see that investment as lower on their list of priorities for spending money [...] but it is great to see where you lie in the whole scheme of things.
>
> ([1] in Table 4.1)

At the same time, Brookdale had also taken the bold step of recruiting a corporate data professional into a newly created role of the school's "Data Analyst". Kyra had previously worked as a business intelligence and data warehousing consultant in various corporate positions. As far as Brookdale's leadership was concerned, this decision to move beyond teacher-led data work to a full-time "professional" hire marked a significant shift:

> We've got to make [data] work for us and it really wasn't happening [...] we've probably been saying that for the last six years [...] and so those conversations evolved into the need of someone like Kyra – but it's not easy to find a data analyst! [...] She's got that technical know-how [...] she's a great resource – we'd hate to lose her.
>
> (Andrejevic, 2019)

Despite her know-how, Kyra had yet to become involved in any data analysis or sustained data-work in the school. Most teachers in the school were only vaguely aware of who she was – as one teacher put it: "So now apparently we have a data analyst. I don't know if I've ever met her. I mean, I've probably said, 'Hello' [...]

don't know what she's used for" ([4] in Table 4.1). Instead, Les had co-opted Kyra into assisting with the redevelopment of the school's IT system ("because she's got fair bit of an IT bent she's had to try and fix [the portals] up from the back end [...] the data part of her job has been put on the backburner" ([3] in Table 4.1)). As Kyra described it:

> I'm supporting the IT team – they need the extra support. My role was data analysis, but I haven't still got to that stage as yet because of the change of products [...] I didn't realise there are so many things in the education sector coming from corporate. [...] so as such [the school] is in a transition phase [...] So I haven't done much data analysis as yet because of this change of products.
>
> (Jarke and Breiter, 2019)

Individual data "enthusiasts"

Stephen, Tony and Kyra were not the only members of staff involved with "doing data". Alongside each school's officially mandated "analysts" were small numbers of highly interested and data savvy staff who also took it upon themselves to engage in data analyses. For example, Brookdale's data leadership were rather closed off from the staff, yet there were a number of lone wolf data analysts doing their own work in the school. In particular, many teachers we spoke with identified Dane as a leading proponent of data in the school. Dane's fellow Religious Education (RE) teachers referred to him as "the master of data at our school [...] he doesn't sleep". Other staff were quick to inform us that "Dane is the data king [...] with his spreadsheets and huge data[sets]" ([12] in Table 4.1). Even the school Principal pointed to "the RE guys" as the main exemplar within Brookdale of "high-end manipulation" ([2] in Table 4.1).

In contrast, Dane was a relatively unassuming young teacher in the RE department who had worked previously as a maths specialist. As Dane explained, what had started as a hobby had progressed in a focus on making frequent use of data within his own teaching:

> I come at it from a statistical perspective. And in terms of my actual expertise and training, it's mostly self-taught. I really enjoy spread sheets. I think there's a lot that I can do with them in terms of figuring out trends and also trying to get a sense of where everything sits. But also just trying to find a half-way house between pure data and the usefulness of that data.
>
> ([13] in Table 4.1)

For Dane, school-related data analysis was now a "passion project". Previously, he had analysed data as a hobby *outside* work, even contributing to an online forum, specifically a subreddit, dedicated to data management ("I used to be fairly active on a subreddit. I would enjoy reading other people's data management problems and then offering solutions on my experience" ([13] in Table 4.1)), he now brought his interests to bear on school-related data problems. Dane would request access to various datasets from the school systems and then set about running analyses that he saw as worthwhile, or otherwise had been suggested by colleagues. As such, Dane has assumed a role as unofficial data consultant for the colleagues he worked with closely, often undertaking analyses that he reckoned could take up much of his evening and weekends. Aside from his love of data, Dane's primary motivation was to use data to "find out what's really happening", "prove points" and "win arguments" within the school. As he put it, "when things get thrown up, I challenge them on a statistical basis if I have to" ([13] in Table 4.1).

Due to the structure of the school's faculties, Dane's efforts remained the preserve of RE and humanities classes. As a relatively traditional Catholic school, Brookdale had a rigid department structure with staff from different subject areas tending to work in "silos". In terms of student numbers and perceived prestige, the RE department was considered to be a leading area of teaching. Moreover, Dane implied that other departments were resistant to sharing good practice, and perhaps also distrustful of his analyses. For example, he bemoaned that "I don't have the credibility" to work with data relating to maths teaching: "[...] there's some very strong characters in the maths department who stand in the way" ([13] in Table 4.1). Dane's colleagues also recalled rebuffed attempts to share data work with the science department: "It tends to be a little bit of 'You do it your way, I'll do it my way'" ([4] in Table 4.1). All told, Dane was keen to "not to do anything too public" and, instead, conduct his analyses for a small (but receptive) group of RE and humanities teachers. As a consequence, teachers in RE and the humanities were using dedicated software to analyse digital data on student learning, while those in the sciences were using more traditional, analogue methods, leading to different faculties having different insights and outcomes when designing future curriculum and pedagogy.

Two similar data enthusiasts were also prominent within Weston – both working within their respective subject "communities" of teachers. Vic was a maths teacher who was involved mostly in working with data generated by a personalised maths learning system being used in most of his department's classes. This generated a mass of data relating to maths teaching that Vic was analysing on behalf of his PLC. Vic was happy to spend considerable time on the "micromanagement of our own data" ([6] in Table 4.1). Vic had begun to share this work with "three other teachers out of the ten or so [in the maths PLC] who want go to this level of detail". Even Tony – Weston's official Data Lead – marvelled at the quantity and specificity of the data that Vic had access to, alongside the amount of time he was able to spend on the analysis: "I've never had enough time to do that with, say, the National Assessment Plan – Literacy and Numeracy (NAPLAN) data" ([9] in Table 4.1).

Elsewhere in Weston, Mick was a charismatic Media Studies teacher who described himself as working with data in a number of "maverick" ways. He described his approach as constituting "almost a new category of teacher that is more research-based as well as pedagogical practice-based" ([14] in Table 4.1).

Mick was certainly confident in pursuing his own agenda. For example, in order to gain better access to data he was opting to circumvent the school's official LMS and run his teaching primarily through Google Classroom. He reasoned that this "hack" allowed him to benefit from what he believed were "seamless" ways of managing student data, assessment and feedback: "it's such a powerful suite [...] happy days!" ([14] in Table 4.1). Mick was also exploring the use of different apps that he thought might generate additional data on students ("I've been messing around the ClassDojo and things like that"). Most notably, and with little regard to privacy, Mick was sharing his data with an informal network of like-minded teachers from other local schools – therefore allowing comparisons to be made with similar students and teaching data:

> We share our data [...] I've got a few networks of teachers, and I've set up things like a shared spreadsheet that looks at both the marks of my class and also some of the other schools in our area. So we're able to cross-moderate, [...] and look at things like the spread of marks in a comparable school. This stuff works really well as a sort of working day-by-day data point [...] People want in, it's not that they have to do it! [laughs] [...] I meet up with one or two of them on the holidays because they're friends, but this is a way of doing collaboration without having to physically meet people.

The "run-of-the-mill" majority

The various staff hierarchies just outlined all dominated how data was talked about within Weston, Brookdale and Northland. Beyond the specialists, however, lay the majority of other teachers. Most staff at each of the three schools were able to identify the "data people" described in the previous sections. Indeed, our initial meeting with Northland's "Data Leadership Team" was prefaced by a clear warning to expect notable differences between "the people in this room and our run-of-the-mill teachers" ([15] in Table 4.1). Other data-focused staff were keen to make similar distinctions between themselves and what were described elsewhere as "average teachers" ([16] in Table 4.1) and those "at the coalface"([2] in Table 4.1).

Of course, many of these "run-of-the-mill" staff were engaging with data in their own ways. As such, there were larger groups of what might be termed dutiful teachers in each school who lacked notable data expertise but nonetheless were well aware of the need to make use of data in their work. Tellingly, in Brookdale this was associated with teachers who have joined the school from the UK: "There's a subset of teachers we've got here[...]who've come from the UK and they are so immersed in data over there [...] It's really interesting hearing them talk about lack of access to data [here]" ([17] in Table 4.1). Often, data was not described by the Australian teachers as a particular passion or interest ("you have to make a special effort, really" ([10] in Table 4.1)), but acknowledged as a requirement of being a contemporary teacher:

> There's quite a few teachers who say that, 'I've been teaching for however many years, don't tell me how to do it.' [...] but there is this now this critical mass in the middle [...] where people are going, 'Come on, get over that!'
>
> ([12] in Table 4.1)

However, an equally large number of teachers were notably less compliant. This antipathy stemmed from a range of familiar concerns and criticisms – for example, relating to the quantification of students: "We tell [students] they are more than a number [...] but then we give them a number" ([18] in Table 4.1). Another teacher who had recently left the profession responded bluntly to the question of how data made her feel: "Bored [...] The data never told me anything new. It just validated what I already knew" ([19] in Table 4.1).

As might be expected, the "data experts" described in previous sections had little sympathy for such attitudes. As one senior leader at Brookdale put it, "there are those who jump on the bandwagon and want to learn about this form of teaching, and others who it leaves behind" ([20] in Table 4.1). Elsewhere, it was suggested these negative reactions to data use were symptomatic of professional incompetence ("some of them are fearful, usually unjustifiably [...] but not always" ([5] in Table 4.1)), generational intransigence ("those who trained in the previous millennium" ([9] in Table 4.1)), or are simply reluctant to be challenged:

> Whatever the data tells you, if it's good and it's true and it's rigorous and it's meaningful, you've got to listen – and that's a hard pill to swallow for a lot of people.
>
> ([14] in Table 4.1)

> Your average teacher follows gut instinct. That instinctive behaviour and learnt pattern of behaviour is not always an easy clear migration to valuing quantitative data.
>
> ([16] in Table 4.1)

Conversely, the increasing importance given to data experts across the three schools was also a source of tacit tension. This was a point of resentment that a few interviewees were happy to voice. On occasion, it was suggested that conspicuous enthusiasm over data in schools was a form of cynical career-building by ambitious colleagues – i.e., "teachers who are looking to build a career and reckon that data can support them [...] they wouldn't think critically or be politically-minded" ([19] in Table 4.1). Elsewhere, the notably gendered nature of many of our previous descriptions was raised as an underlying issue. In justifying Brookdale's data drive, for example, one teacher reasoned simply, "The place is run by men you see" ([12] in Table 4.1). Otherwise, data was rationalised as a matter of status and power: "It's not a matter of gender [...] but it is a matter of hierarchy" ([18] in Table 4.1). All told, leadership assumptions that "the staff here are open to data" were certainly open to debate.

Discussion

In one sense, Weston, Brookdale and Northland could all be described as relatively successful users of data. On the face of it, Principals and senior staff had sufficiently "crafted coherence" (Gunnulfsen and Roe, 2018) around the idea of being "data schools". While not boasting notably distinctive or innovative "data cultures" per se, each school could nevertheless plausibly claim to be paying attention to student and teacher data. Committees and teams had been convened, individual members of staff appointed in "lead" and "manager" positions, and teachers were generally assumed to be "on board". All told, these schools were certainly making "good enough" use of data without committing fully to data-driven and data-intensive practices.

While some broad similarities were apparent, our findings also point to the influence of local school contexts on the ways in which data was being approached and enacted. For example, Weston was a relatively open and progressively-minded school – subject to the state government's data "imperatives", but also striving to foster a collective approach in the way that it went about things. In contrast, Brookdale was a relatively under-resourced traditional Catholic school – lacking the same centralised imperatives but nevertheless feeling compelled to "do data". Finally, Northland was a fee-paying school with a distinct corporate ethos and focus on efficiency. These different values and agendas certainly shaped each school's approach towards data. At the same time, we need to acknowledge the influence of school micropolitics on the ways in which particular staff were engaging with different forms of data. For example, there were distinct differences in how top-down executive control was (and was not) being challenged, the legacies of long-standing departmental divisions and inter-subject hostilities, as well as how new data-related organisational arrangements were being subsumed into established patterns of power and control.

Interestingly, across all three schools, data was approached as an "object" to be used and valued. Data is "operational" in that it is ostensibly strings of indecipherable numbers that flow between software systems (Andrejevic, 2019), however, it was not often seen as such. This has ramifications for how data becomes entangled with power. As an object, digital data can be traded, moved and correlated – it becomes valuable to those with the knowledge and expertise to "do something with it", regardless of how useful or innovative that might be. However, in reality it is often how software systems process and represent the data that is significant. Shifting away from the data as "object" perspective might lead to a more critical and holistic perspective on how education *systems* are changing due to datafication and the commercialisation of educational technology. As Hintz et al. (2019) explain, datafication "facilitates a new form of governance that relies on data-driven practices of categorization, classification, segmentation, selection and scoring" (p.146). What we observed in each of our schools is the datafication of systems and processes, yet this was often not discussed. Instead, data was considered an object that was to be dealt with by particular people and/or groups in each school.

The growing prominence of data in these three schools was also aligned with a number of notable data-related reconfigurations of power and redistributions of agency (Jarke and Breiter, 2019). First, despite appearing to be institution-wide commitments, only a handful of staff were actually involved in each school's data-work. Beyond each school's executive oversight and "data leadership" group, actual data work remained the preserve of a few staff with sufficient data skills and interest, but different responsibilities and accountabilities. In this sense, the establishment of data groups, data deans and project managers were largely organisational in effect – "the establishment of bureaucracy as a purpose of, rather than as a means for, technological advancement" (Wittmann, 2018, p.70). More significant, then, were the individuals actually engaging with school data on a sustained basis – i.e., either in the guise of officially mandated "data leads" or as self-motivated "data mavericks".

On the face of it, the creation of official "data lead" roles reflect a tendency for schools to respond to technological change by creating additional specialisations and positions (cf. the recent rise of school CTOs, network managers and "digital innovation" leads). While teachers such as Stephen and Tony were self-proclaimed data "enthusiasts", their designated "data lead" roles were clearly demanding. These positions involved intense bouts of data-work in response to requests from school leadership – therefore constituting what could be described as quasi-consultancy roles. With this remit came a considerable amount of "soft power", which could be used to coerce, highlight or nudge particular behaviours in teachers and students. This was evident in the multiple spreadsheet-based decisions that each teacher made about what could (and could not) be said about the school, teachers and students. These analytic "tweaks" and "fiddlings" were being made on the basis of Stephen and Tony's technical and professional judgements – not least, their personal readings of the datasets and tacit assumptions about the students and staff represented in the data. In this sense, both Stephen and Tony were key gatekeepers of "what the data said", and therefore held undefined but influential positions of power within their schools (see Wood, 2014).

In contrast, data mavericks such as Mick, Dane and Vic were not beholden to the analytical demands of their superiors. These staff presented themselves as striving to use data as a reality check on their own teaching and the teaching of close colleagues. While these staff also considered data-work as a hobby and passion, they were clearly following their own lines of interest and inquiry (and were therefore less interested in fostering any schoolwide use of data). These individuals could be said to be exercising power within their personal networks – i.e., using data to gather the respect and admiration of peers within and beyond the school. Yet, these teachers' data work was not being undertaken to direct the actions of others per se, but rather being used to build their own status and careers, and in some cases, to be of benefit for their colleagues.

For the time being, at least, these two cadres of data analysts loomed large over how their schools were engaging with data. The rising significance of such staff reflects the emphasis currently attached to the specialist "data analyst" in the practical implementation of the data-driven society. These are individuals that promise to tame and draw meaning from masses of data. Our schools were therefore understandably happy to delegate some responsibility for "doing data" to teachers willing to act-out the analyst role – hopeful that they might coax some extra sense and additional narrative from their data. As David Beer reflects:

> These figures are expected to create insights through a combination of skill, analytical ability and the capability to know and martial both the data and the analytical tools available. Time pressed and data rich, the analyst's skilled glance is seen to be of great value.
>
> (Beer, 2019)

As such, these data analysts were prominent role-models for the growing impingement of data on reformations of teacher identity and teacher professionalism – i.e., "educators' understanding of themselves as professionals" (Anderson and Cohen, 2015, p.2). Indeed, the pressure of "data" has already been noted as a significant factor now shaping teachers' professional identities (Lewis and Holloway 2013; Buchanan and McPherson, 2019). In our own study, the imperative of data and the visibility of school data "leads" and "mavericks" was also an acknowledged influence on how most teachers saw themselves – either in terms of obediently "getting up to speed" with data at some time in the near-future, or else a recalcitrance to be "doing data" at all. That said, the strongest sense of altered professional identity arguably resided in the "data mavericks" themselves. Some of these teachers clearly saw themselves as a "new breed" of professional, and certainly were drawing on their use of data to reposition themselves amongst colleagues. Teachers such as Mick could be said to embody what has been noted elsewhere as a "new professionalism" within teaching – i.e., driven fewer by issues of discretion, trust and partnership, and more towards "outcomes-based, entrepreneurial and corporate" ways of working (Anderson and Cohen, 2015, p.3).

Indeed, this perceived distinctiveness of "doing data" could be said to correspond with at least three differentiated identities or forms of "capital" that were distributed unevenly amongst staff in our three schools. First is what might be called "data capital" or "analytic capital" – i.e., the distinction of having "huge data" or simply being able to talk comfortably about pivot tables. Throughout all our time in the schools, the issue of actual competence in data analysis was a carefully choreographed topic of conversation. The older official "data leads" in Weston and Northland were keen to underplay their analytic skills. In contrast, Brookdale's business intelligence professional was diplomatic enough not to appear critical of the "small data" and "primitive tools" that she was encountering. Even the "data mavericks" were not excessively boastful. Indeed, as Mick reflected: "I think no-one in education is an expert in data, no matter what they tell you – no-one comes in as being a top-end statistician" (Mandinach, 2012). As such, the ability to maintain an appearance of analytic competency without over-claiming one's skillset was a key capacity. In this sense, "data power" in schools was not so much derived from advanced statistical know-how, but the capacity to project a general confidence and competence with data.

Allied with this was an associated "technical capital" (Sims, 2014). This was key distinguishing feature between the school "data leads" (notable for their "old-fashioned" reliance on Excel), and the more technologically ambitious "data mavericks". As has been noted before, digital skills and technological expertise are often used by individual teachers as "a means to strengthen their position within their school" (Pitzalis and De, 2019, p.85). Crucially, the data mavericks were drawing on a wide range of technological resources - from the transgressive adoption of platforms such as Google Class through to drawing on forums such as Reddit and social-media networks of teachers from other schools. Compared with the relatively staid nature of their schools' technology use, these teachers were notably engaged in mainstream "tech culture" and networked practices – a key part of how they distinguished themselves from their colleagues. In this small way, then, "demonstrating resistance can generate symbolic capital to be invested and exploited as a career furthering strategy" (Pitzalis and De, 2019, p.86).

Finally, alongside these technical capitals was the continued significance of what might be termed "teacher capital" (Zevenbergen, 2006). "Data mavericks" such as Mick, Dane and Vic were well-regarded by their immediate colleagues for deploying advanced uses of technology and data to improve teaching practice and better understand their students. In contrast, Brookdale's "great resource" of hiring Kyra was treated with distinct ambivalence by most teachers we spoke with – despite her extensive skills as a corporate business intelligence analyst. This notion of teacher capital was also apparent in the resistant refrain of "knowing" students as more than a number. In these ways, the kudos of such "teacherly" concerns remained a key element of who was permitted to be a valid school data expert and/or why some teachers were keen to present themselves as disengaged from data use.

Indeed, the various forms of data work just described appear to be delineated by a range of notable factors – not least the gender, status, subject knowledge and career stage of the staff involved. These characteristics were also apparent in the types of teachers more likely to be marginalised and excluded from the beneficial consequences of "doing data". In particular, our study points to a number of ways in which in the "doing" of school data was arranged along gendered lines – i.e., the ways in which gender and identity was being constructed in and through school data, as well as how data expertise and engagement was situated, embodied and localised in the marginalised voices and views of less powerful female staff. For example, data analyst Kyra had taken the professional step backwards to work in schools to fit with her childcare commitments, while the male data specialists and mavericks were able to work long hours on the weekends and in the evenings given their lack of immediate caring responsibilities. Such instances highlight ways in which data inequalities map onto existing power differentials and unequal social relations across school staff. As Evelyn Ruppert (2013) points out, it is notable how the dominant "datascapes" of societal settings such as schools tend to be tied closely with dominant "theories of social order".

These findings point to several broader paradigm shifts emerging within schooling systems. Not only is there a prioritising of "objective" judgement of student performance, but this "new professionalism" is changing classroom practice as teachers begin to prioritise activities and assessments that can be easily datafied, as opposed to those that are best for student learning. For example, Mick was exploring how he could use different software programs like Class Dojo and Google Classroom to generate more data on his students so he could compare their performance with students at other schools. In this way, teachers and students are becoming subject to datafication processes. They are, as Ajunwa and Greene (2019) explain, 'supervised' by software. In critical circles, datafication raises concerns of privacy and profiling, yet our findings highlight a more nuanced – but equally problematic – set of concerns specific to the school context.

While issues of privacy and profiling are important, datafication was leading to the reconfiguring of teacher roles and schooling processes before many in the school had a chance to consider exactly what was taking place. As such, it was technology, in combination with the priorities of certain data classes, that were driving the agenda, while everyone else simply had to "get on board". Our findings point to the need for *all* teachers to be better supported to develop understandings of data and datafication. Only then is it possible to identify and potentially resist how the datafication of education is changing classroom practice. Curriculum and pedagogy must be designed and implemented by teachers with content *and* context knowledge (i.e., students, resources), in order to serve the most important stakeholder – the student. The power of 'big tech' and data processing may serve a bureaucratic purpose, but it will be potentially damaging if it is directing the everyday teaching and learning experiences in schools.

Conclusions

In some ways, these findings could be read simply as schools struggling to satisfactorily respond to external imperatives to "do data". Clearly, we did not find schoolwide engagement with data – instead, small groups of staff continue to drive their school's interest in data and take effective ownership of the "issue". Instead of criticising Northland, Weston and Brookdale as not fitting the mould of the ideal-type "data-driven" school, all three schools might in fact be praised for resisting the feared repercussions of excessive data-driven performativity and professional disempowerment. For the time being, at least, these schools were certainly not being over-run by data.

In this sense, the current ad-hoc arrangements described in this chapter need to be made "problematic" – i.e., framed as a matter of school politics, and reconsidered as a topic of professional concern and controversy. In this sense, there is much in this chapter that needs to be challenged. For example, how might the concerns of run-of-the-mill majorities be better represented in how schools collectively view their engagements with data? How might the topic of school data be reconstituted as a more collective, democratic schoolwide endeavour? What alternate forms of data expertise and data uses might be encouraged within the school communities? Rather than providing definite answers to educational problems, the use of data within schools clearly needs to be approached in a more sceptical way.

TABLE 4.1 List of staff participants

	Surname	*First name*	*Role*	*F/M*	*Age*
1	Weston	Sarah	Principal	F	51–60
2	Brookdale	Stewart	Deputy (one of three)	M	51–60
3	Brookdale	Les	Dean of Data	M	61–65
4	Brookdale	John	RE teacher	M	31–35
5	Northland	Dennis	Principal	M	51–60
6	Weston	Vic	Maths teacher, PLC leader	M	51–60
7	Weston	Susan	Assistant Principal	F	51–60
8	Northland	Stephen	Data lead teacher, maths	M	51–60
9	Weston	Tony	Data lead teacher, biology	M	61–65
10	Weston	Ben	PE teacher, PLC leader	M	21–30
11	Brookdale	Kyra	Data analyst	F	31–40
12	Brookdale	Larissa	Teaching librarian	F	65+
13	Brookdale	Dane	RE teacher, data lead for subject	M	21–30
14	Weston	Mick	English teacher, PLC leader	M	31–40
15	Northland	Russ	Deputy Principal	M	31–40
16	Northland	Rob	Senior IT technician	M	41–50
17	Brookdale	Richard	Humanities teacher, Head of Learning and Teaching	M	51–60
18	Northland	Angela	Head of English literature and drama	F	41–50
19	—	Jo	Ex-teacher – left profession at the end of last academic year	F	21–30
20	Brookdale	Christian	Humanities teacher, LMS manager	M	31–40

References

Ajunwa, I. and Greene, D. (2019). Platforms at work. in Vallas, S. and Kovalainen, A. (Eds.) *Work and labor in the digital age*. Emerald (pp. 61–91).

Anderson, G. and Cohen, M. (2015). Redesigning the identities of teachers and leaders. *Education Policy Analysis Archives*, *23*(85): 1–28.

Andrejevic, M. (2019). *Automated media*. Abingdon Oxon, UK, Routledge.

Ball, S. (1987). *The micropolitics of the school*. London, Methuen.

Beer, D. (2019). *The data gaze*. London, Sage.

Buchanan, R. and McPherson, A. (2019) Teachers and learners in a time of big data. *Journal of Philosophy in Schools*, 6(1) www.ojs.unisa.edu.au/index.php/jps/article/view/1566.

Gunnulfsen, A. and Roe, A. (2018). Investigating teachers' and school principals' enactments of national testing policies. *Journal of Educational Administration*, *56*(3), 332–349.

Hintz, A., Dencik, L., and Wahl-Jorgensen, K. (2019). *Digital Citizenship in a Datafied Society*. Cambridge, UK, Polity Press.

Iliadis, A., and Russo, F. (2016). Critical data studies: An introduction. *Big Data & Society 3*(2) doi: 2053951716674238.

Jarke, J., and Breiter, A. (2019). Editorial: The datafication of education, *Learning, Media and Technology*, *44*(1):1–6.

Kennedy, H., and Bates, J. (2017). Data power in material contexts: Introduction. *Television & New Media*, *18*(8), 701–705.

Kim, K. (2018). Early childhood teachers' work and technology in an era of assessment, *European Early Childhood Education Research Journal, 26*(6): 927–939.

Mandinach, E. (2012). A perfect time for data use: using data- driven decision-making to inform practice, *Educational Psychologist, 47*(2): 71–85.

Mandinach, E. and Gummer, E. (2013). A systemic view of implementing data literacy in educator preparation. *Educational Researcher 42*(1): 30–37.

Pitzalis, M. and De Feo, A. (2019). Micropolitics of school innovation. *Italian Journal of Sociology of Education, 11*(1): 69–90.

Schildkamp, K., Poortman, C., Luyten, H. and Ebbeler, J. (2017). Factors promoting and hindering data-based decision-making in schools, *School Effectiveness and School Improvement, 28*(2): 242–258.

Sims, C. (2014). From differentiated use to differentiating practices. *Information, Communication & Society, 17*(6): 670–682.

Wardrip, P. and Herman, P. (2018). 'We're keeping on top of the students': Making sense of test data with more informal data in a grade-level instructional team, *Teacher Development, 22*(1): 31–50.

Wittmann, E. (2018). Serving the individual or bureaucracy in educational administration? in *Samier, E.* and Milley, P. (eds) *International perspectives on maladministration in education*. London: Routledge (pp. 63–74).

Wood, P. (2014). *Teacher professionalism*. Oxford, Symposium.

Zevenbergen, R. (2006). Teacher identity from a Bourdieuian perspective. in Grootenboer, P. (ed.) *Identities, cultures and learning spaces*. Sydney, MERGA (pp. 616–618).

5

EDUCATORS' DATA LITERACY

Understanding the bigger picture

Juliana E. Raffaghelli

Introduction

Data extraction and algorithmic manipulation have become increasingly frequent across the globe. Initially perceived as an opportunity to innovate in several areas of human knowledge (Stephen-Davidowitz, 2017), data practices had unthinkable impacts. Race, gender and other vulnerable characteristics were made invisible, overrepresented and over-tracked by certain forms of data visualisation (Ricaurte, 2019). Automations based on machine learning have entailed perilous biases in the way they represent or neglect relevant cultural perspectives (Malik, 2020). Moreover, they have led to biases and users' lack of agency or even harm (Eubanks, 2018; O'Neil, 2016). In education, research has also highlighted forms of surveillance on children, teenagers and young adults with the aim of addressing their behaviours, emotions and cognition (Chi et al., 2018; Lupton and Williamson, 2017; Prinsloo, 2020). The "platformisation" of learning and the monetisation of students' data are also pressing issues in the educational agenda. These problems were magnified by the intense and unprecedented use of digital technologies during the pandemic (Perrotta et al., 2020; Williamson et al., 2020).

Against this backdrop, an activist counterculture emerged (Baack, 2015) in which participants aim to uncover how surveillance is redistributing power through the participatory appropriation of data (Lehtiniemi and Ruckenstein, 2019). Expressions of this movement have manifest through the Open Government Data and Open Science (Lehtiniemi and Ruckenstein, 2019), as well as with independent collectives reacting to the oppression of surveillance, including forms of disconnection and "hacking" the system (Pybus et al., 2015). Overall, data activists' search for recognition, awareness, redistribution of symbolic and material power aims at emancipation. In light of this, there is not a universal approach to data literacy that institutions and educational systems can embrace.

DOI: 10.4324/9781003136842-5

In this chapter, I contend that there is a need for educators to develop a more complex and critical perspective on data in society and in education. I argue that educators and institutions should engage in explorative approaches to unveil diversified dispositions to data. To explore this approach, I conducted 12 workshops (eight with English speaking participants, three with the Spanish speaking community and one in the Italian context) from 2018 to 2020. Overall, nearly 294 educators, PhD students and faculty professors participated. The methodology was held constant along the several workshops: I introduced several data practices in education and society, to trigger reflection about the educators' professional knowledge and practice. The workshops involved hermeneutic circles where the conversation between myself, as the researcher, and the participants triggered understanding around educators' literacies to deal with situated data narratives and practices. This conversation unveiled the tensions between dispositions to data (reactive or proactive) and the forms of access to data (public or private), as spaces where the educators intervene. The data literacies required are diversified across these spaces, but they are also part of an integrated mosaic that can vary from one context to another. My ultimate consideration is that faculty, students and other stakeholders' efforts to develop data policies, strategies and practices, should address the larger challenges of datafication in society.

Conceptual lens

As Milan and van der Velden (2016) argue, data circulates generating imaginaries, discourses and practices, which they refer to as "data epistemologies". Data epistemologies focus on the contextual and diversified nature of data practices in response or reaction to datafication. According to Milan and van der Velden, the activists that search for data appropriation to express diversity and empowerment embrace a *proactive* data epistemology. While the activists whose goal is to uncover injustice and condemn malpractice adopt a *reactive* data epistemology. In each case, the quest for equal representation, voice and self-determination leads to an ideal known as fair data culture.

From an educational perspective, the idea of "data epistemologies" needs to be translated into an "actionable" concept that allows the beliefs, attitudes and unconscious understandings of data that underpin human activity and educators' practices to be developed. In this regard, I adopt the term "dispositions towards data", to signal the attitudes and beliefs people have towards data, which incline them to particular ways of reacting to it. Indeed, dispositions can be seen on a continuum, in which extremes are represented as either proactive or reactive ideas, practices, discourses and responses to data. Nonetheless, transforming both reactive and proactive dispositions towards data into action requires advanced technical skills and civic engagement that go well beyond the actual possibilities marginalised and vulnerable collectives may have.

Education plays a crucial role by developing data literacies that move between proactive and reactive dispositions. Moreover, educators' awareness of their own dispositions towards data might help them transform the complexity of data into their pedagogical practice. This statement, however, has immediate implications: if education has to respond to the context of innovation and continuous transformation of data practices, and if several data approaches can be embraced, the educator's response could become fragmented. As a result, it would be difficult to promote creative and agentic ways of relating to data in education and the society (Raffaghelli, 2020b). In fact, the initial research around data literacy has focused a narrow and instrumental conception, which does not promote complexity in data practices (Raffaghelli and Stewart, 2020).

Promoting complex understandings of data in society requires embracing approaches that support learners to become aware of and actively take control over data practices, both as individuals and collectives. (Markham, 2018). While it is clear that technical literacies are relevant to read, manage, process and visualise data, and interact with algorithms and develop a proactive disposition to data, critical literacies around a participatory and dialogic culture explore meaning, power and the material conditions of data practices developing reactive disposition to data. In this sense, a complex understanding of data literacy goes beyond the technical and the critical component and encompasses the integration of the data approaches. In the following, I will consider some theories and concepts helping us to understand the evolution of data literacy as a research topic.

Paulo Freire's critical pedagogy has already tackled the need to integrate the technical and the socio-cultural layers (Freire, 2000). Through his theoretical lens, the transfer of knowledge from the elite to the community builds contextual knowledge by appropriation of the technical means in order to represent such world. For Freire, these means were reading and writing. Later on, the digital society shifted the debate on the digital medium and the needed skills. While there has been strong attention to the technical skills required to be digitally literate, the ability, knowledge and dispositions to live in a digital society, are also necessary. Using digital technologies critically also requires some degree of designing and creativity (Pangrazio, 2016). In short, technical and critical skills are important to engaging with the digital medium in a way that is both safe and creative (Tour et al., 2021), Moreover, as Pötzsch points out, digital literacies need to move beyond responsible and efficient digital technology use and focus on 'capitalist dynamics, environmental ramifications, and individual empowerment' (2019, p. 222).

Nonetheless, in recent years much emphasis has been placed on coding (Popat and Starkey, 2019), the maker movement (Papavlasopoulou et al., 2017) and digital competence (Carretero et al., 2017). Initial claims for the potential of digital competence to be part of a rapidly changing and smart society turned later into a series of predictable difficulties in promoting access or socially relevant learning, beyond the technological skills (Sancho-Gil et al., 2020).

Similarly, the debate around data literacy focused first on the technical dimension (Raffaghelli and Stewart, 2020). Discussions on numeracy and then on statistical literacy paved the way for a technical perspective on data literacy (Gould, 2017). This strand of research addressed the technical comprehension of algorithms, data-driven practices, data visualisation and programming as the creative side of data science (Gray et al., 2018; Wilkerson and Laina, 2017). Aligning with this trend, the digital tools to process and visualise data, as well as the opportunities for dynamic representations, recommendations or activities, opened an area of skill development in higher education. Overall, data literacy mainly covered individual practices within the classroom, both from the side of data as educational content and data as part of inquiry tasks (Maybee and Zilinski, 2015).

Nevertheless, the aggregation and analysis of data gathered from assessment and processed as quality evaluation metrics can initiate reflection, with data considered as an element in support of pedagogical methods (Mandinach and Gummer, 2016). Less emphasis has been put on understanding how students can be empowered by the use of data within the classroom and towards their communities (Raffaghelli, 2020a), though the debate around students' data usage is evolving at a steady pace (Willis et al., 2016). More recently, ideas such as data justice and data feminism have permeated the educational system in search of broader perspectives on data for the society and the educational institutions themselves (D'Ignazio and Bhargava, 2015; Raffaghelli, 2020b). Nonetheless, authors have advanced the need of approaching data literacy through a critical lens (Pangrazio and Selwyn, 2019), as well as through the adoption of Freire's perspective (Manca et al., 2017; Tygel and Kirsch, 2016).

Moving beyond the technical approach, critical data literacies can also explore different data practices: from promoting open data appropriation in formal and informal contexts of learning (Baack, 2015; Manca et al., 2017) to considering personal data privacy (Pangrazio and Selwyn, 2020), awareness of data usage and algorithmic manipulation across digital learning platforms (Stewart, 2020). The literature considered here highlights tensions between different dispositions towards data re-emerges and produces entropy in the way the educational intervention might be characterised and implemented. The liaisons between the several data usage scenarios and which approach to take is left to the educator who is often in professional isolation. To strike a balance between the proactive and reactive dispositions towards data mentioned in the introduction, the emerging data practices should be identified and conceptualised to support educators' reflection and organic action, as well as to promote critical data literacies through a complex integrated data literacy's "bigger picture". I will now introduce my approach, based on building professional learning spaces to reflect on the complexity of data in society and within educational practice.

Methodological approach: a hermeneutic journey to explore the boundaries of data literacy with and for educators

Context of implementation and research goals

In the context of a five-year (2018–23) national project supporting faculty development in changing digital contexts[1], my overall research work aimed at exploring and mapping data practices in higher education institutions (first and second year) as a means to build a framework (third year) that would be helpful for educators to promote critical data literacies. During the initial phases of the project, the map of data practices proved to be fragmented, mostly connected to a discourse of innovation (i.e., learning analytics) and to open science (i.e., open data) (Raffaghelli, 2020b). For the scholars and the institutions I interacted with, data was seen as something happening in the present but continuously evolving to shape either a utopic or dystopic future: If we do not catch up data-driven practices, will we step outside a trend of necessary innovations to promote new skills and new jobs? Will data practices produce more harm than good? Based on an interdisciplinary approach, the project promoted a vision of data literacy that included technologies, politics, aesthetics and narratives of data, searching for social justice as the ultimate goal of any educational action (Raffaghelli, 2018).

In this context, I carried out a number of workshops for educators, as part of an initial exploratory activity to understand what actions were needed to build educational practices focusing on data in society. The educators and educational researchers' specific goal was to identify the impact of emerging data discourses and practices in connection with their professional identity. Most of the workshops were conducted during international conferences, though some were part of faculty development strategies.

Research and educational design

From a methodological point of view, the workshops could be characterised as nodes in an iterative action conducted by the researcher with educators' engagement, aimed at building a bigger, integrated picture around data practices and what literacies might be needed. For this purpose, I drew on the philosophical concept of the "hermeneutic circle" (Gadamer, 1960/2013). The German philosopher conceived it as "a movement" in "repeated return from the whole to the parts and vice versa" (p. 196), between the reader and text, within conversation. Hence, it involves a dialogue with the world and others (Gadamer, 1960/2013, p. 386) and refers to a situated knowledge, prompting a contextualised search of truth, which cannot be reduced to a set of research methods. This approach has already been embraced in educational research (i.e., in-depth interpretation, jointly with the learners and their communities) leading educators to frame educational problems, design for learning and search for adjustments between the educational act and its impact on the reality (Carr, 2007). In this study, I developed a workshop design with activities and a conceptual tool (listed below) to think on data practices. This then became the participants' object of critique and enrichment. Engagement elicited feedback and challenged our (educators' and mine) understanding of data literacy as a whole, exploring the spaces for educational intervention. In this regard, the sampling was theoretical, not based on the participants and their responses, but on the categories of analysis' saturation (Corbin and Strauss, 2007).

Since the topics tackled different data practices, the workshop structure as a hermeneutic-pedagogical approach was constant and constituted of:

- **Initial information** to generate a sense of curiosity.
- **Self-diagnosis** to capture insights and develop self-awareness.

- **Technical instruments to engage with data as material expression**, based on methods, jargon, technologies, practices and knowledge within professional communities.
- **Reflection** connected with the participants' experiences and practice contexts based on the epistemologies of data.
- What am I doing and why am I not working with data? Why am I engaged in this movement, or why am I not? Am I resisting the establishment, or am I trying to survive within an unjust context of professional practice and personal life?
- **Exploration of the professional roles as educators and educational researchers**: *'If I behave in a certain way, to what extent am I providing a different vision to my students or to the participants of my educational research?'*
- **Representation of the reflection's outcomes through a quadrant graph as a conceptual tool.** Elaborating on each of the conversations with the educators, I modified the pair of tensions forming the quadrant. The dimensions generated spaces to help categorise the educators' fragmented, lived, perhaps untold, experience.

Figure 5.1 introduces the initial quadrant graph of "data practices and literacies", being the goal to improve the categories and to characterise the types of data literacies needed to move across the four quadrants. In each of the four quadrants, spaces are delimited by the tension between access to data/knowledge and the educators'/learners' dispositions toward data. The theoretical foundations considered public

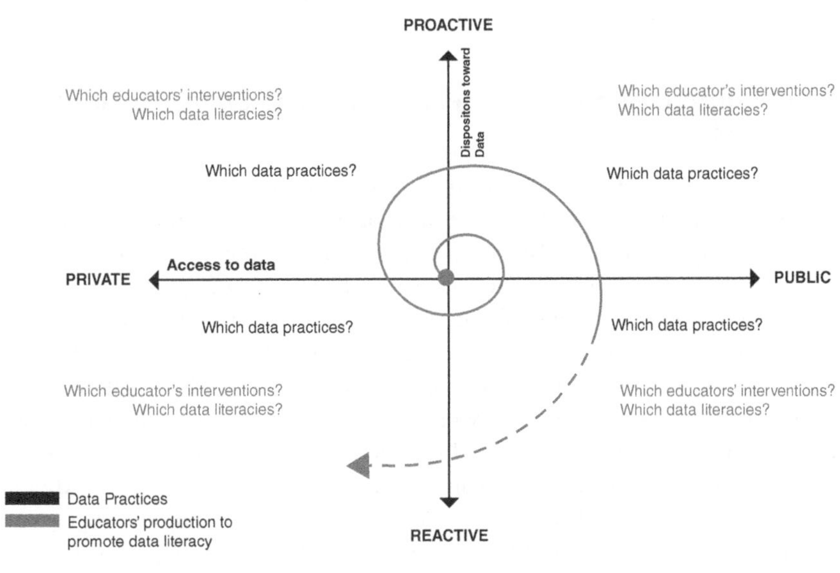

FIGURE 5.1 The quadrant graph as conceptual tool to accommodate educators' perspectives on data literacy around data practices.

data access and the open data movement, as well as private data access and the concerns of data surveillance and algorithmic manipulation. During the workshops, my aim was to place a series of specific data practices within each area, followed by the required educational practices to promote data literacy. A recursively complex picture around data practices and literacies could have been achieved through the educators' participation, reflection and understanding across the four fields.

Table 5.1 shows the workshops' names and link to the published open resources and reports. Each workshop could be considered the beginning and the end of a hermeneutic circle. The table also includes the number of participants, the working languages, the participants' expertise, the expected targets, the number of encounters and duration. Also, the geographical areas covered can be found in the seventh column. All the mentioned columns portray the extension and diversity of education professionals who participated and engaged in the activity. A breakdown point generated by the beginning of the pandemic was noted, so I moved the organisation of the workshops to online activities.

The most relevant element, that is, the workshop focus on data practices, is reported in the "data approach" column. The data approach shaped the topics dealt with and the types of resources and activities (though the methodology was constant). In that regard, I selected the workshop topics over the prior hermeneutic iteration.

The data approaches adopted in each workshop were:

- A= Open Research Data for Open Educational Science.
- B= Open Data as Open Educational Resources.
- C= Data Surveillance in Education and Society.
- D= Data for Good in Education.

Results: Interpreting and understanding data practices with the educators

As observed in Table 5.1, data practices were concentrated mainly in the initial workshops (W1 to W4), which began with open data in educational research and moved to the use of open data as open educational resources, sticking to epistemologies of proactive data. The educators' participation and questions were oriented towards the exploration of ethical concerns and possible implications for the educational practice from the perspective of epistemologies of reactive data. The starting point was based on data practices in education that I had explored in my prior research. However, I purposely provided an incomplete structure with an aim to trigger the hermeneutic cycle and complete the quadrant graph as a final representation of the collective understanding of educators' engagement with data literacy in their professional activities. In the succession of workshops, I either invited experts or collaborated with colleagues whose expertise would support participants' reflections while I led the workshops' methodological approach. In the following, I

TABLE 5.1 Implemented Workshops

Workshop's Name	*Sched.*	*N*	*Lang*	*Expertise*	*Targets*	*Geographical Distribution*	*Time*	*M*	*Data Approach*	*Mode*	*Report*
W1/ NWL 2018	May 2018	9	ENG	Advanced	Educational Researchers Higher Education	Europe	60	1	A	onsite	http://doi. org/10.5281/ zenodo.4446013
W2/EDEN 18	June 2018	8	ENG	Advanced	Transnational Projects' LeadersEducational ResearchersVET educatorsOpen Education promoters	Europe	90	1	B	onsite	http://doi. org/10.5281/ zenodo.4446267
W3/ MASTER UB19	Feb 2019	16	SPA	Mid-career Advanced	Teachers' Educators, Educational Technologists	Europe Nord America Latin America	240	2	B	onsite	http://doi. org/10.5281/ zenodo.4446763
W4/JTELSS 19	June 2019	25	ENG	Initial	PhD candidates in technology-enhanced learning	Europe	60	1	A	onsite	http://doi. org/10.5281/ zenodo.4446844
W5/OEG19	Nov 2019	6	ENG	Mid-career Advanced	Transnational Projects' LeadersEducational ResearchersOpen Education promotersHigher Education experts	EuropeNord America	60	1	A-B	onsite	http://doi. org/10.5281/ zenodo.3557431
W6/UOC 2019	Dec 2019	12	ENG	Initial	PhD candidates Research Assistantsin technology-enhanced learning	EuropeNord America	180	1	C	onsite	http://doi. org/10.5281/ zenodo.4453078
W7/UBAAP 2020	Jan 2020	10	SPA	Mid-career Advanced	Higher Education Teachers	Catalonia-Spain (Europe)	180	1	D	blended	http://doi. org/10.5281/ zenodo.4460680

(Continued)

TABLE 5.1 (Continued)

Workshop's Name	*Sched.*	*N*	*Lang*	*Expertise*	*Targets*	*Geographical Distribution*	*Time*	*M*	*Data Approach*	*Mode*	*Report*
W8/OER 2020.I	April 2020	73	ENG	Mid-career Advanced	Transnational Projects' LeadersEducational ResearchersOpen Education promotersHigher Education experts	Europe Nord America Latin America	30	1	B	online	http://doi.org/10.5281/zenodo.3739180
W9/OER 2020.II	April 2020	94	ENG	Mid-career Advanced	Transnational Projects' LeadersEducational ResearchersOpen Education promotersHigher Education experts	EuropeNord America	60	1	C	online	http://doi.org/10.5281/zenodo.3744135
W10/AREA 2020	April 2020	25	SPA	Mid-career Advanced	Educational technologistsOpen Education promotersHigher Education teachersTeachers' Educators	Latin America	90	1	B	online	http://doi.org/10.5281/zenodo.3839439
W11/MED 2020	Nov 2020	5	ITA	InitialMid-Career	Teachers' Educators, Educational Technologists	Italy (Europe)	240	4	B & C	online	https://www.youtube.com/watch?v=F0Z-BWGPcwQ
W12/LATAM 2020	Dec 2020	15	SPA	All levels	Higher Education Teachers	Latin America	60	2	A to D	online	http://doi.org/10.5281/zenodo.4475965

TOTAL WORKSHOPS: 12 – TOTAL PARTICIPANTS= 298

introduce quotations representing key elements in some of the most relevant hermeneutic iterations.[2]

Initially, I used the quadrants from the first circle in the A–B type workshops to characterise the personal motivation versus the institutional support to become open scholars and share/re-use open data. By exploring the several positions and beliefs around the role of data in society and the educator's profession, I became aware that the dispositions towards data played a crucial role. As it emerged in the W1 (Report W1[3], p. 3), some participants (SJ and PM) were concerned about how data could be extracted whereas MH considered the ethical conundrum in using students' data:

> If a person or student leaves their data on a social network, even if it is public, can it be collected and processed for research? This was not the original purpose.
>
> (MH)

> One bigger problem is the data that could not be accessed due to the obscure forms of data processing by social networks.
>
> (SJ)

> It appears that data, which can be necessary for research purposes (and hence for good, hopefully!), is more and more difficult to get.
>
> (PM)

> ...data is more and more difficult to get (due to ethical concerns and procedures) whereas for the companies, data is easily available.
>
> (SJ)

In the following workshops (W2;W3), the problem of open data extraction or usage within the classroom for teaching purposes was considered by the participants. I continued exploring the right side of the quadrants. Educators held that providing the class with open data as resources, as well as promoting civic engagement and monitoring open data could imply innovation and discovery, hence, massive intellectual efforts (Report W2, p. 8):

> The main issue was to identify which data and which levels of literacy had to be considered: my students would not be able of participating in open hackathons.
>
> (FN)

> The materials should be produced to prompt relevant reflections and engagement with data, whilst also being aware of the risk of failure due to the technical complexity of extracting, managing, elaborating and presenting data (to students).
>
> (SM)

On these grounds, for the hermeneutic circle referring to OEG19 (Report cited in the table, p. 4), I re-proposed the quadrants to explore public vs private knowledge and personal vs group generated data in order to better grasp the expressed problems of access to data. The participants (divided in two groups) engaged with these categories in the following ways:

> Awareness and debate come first (relating data practices)! But we are not sure about the policy making that could come at a late stage.
>
> (Group 1)

> We need training in educational data management and and research data management...know the basics before you dive in deeper.
>
> (Group 1)

However, one participant insisted on the relevance of AI and data: "Awareness and debate to support participatory design for Artificial Intelligence integration will be at the cutting edge" (LH, Group 2). Within this group, the participants concluded indeed that the category of "data analysis and algorithms transparency" was considered transversal to other forms of knowledge and awareness. Again, the different approaches between groups in this workshop highlighted the presence of diversified dispositions to data. The attention in this group was all devoted to the need of making data a public good.

From Workshop 6, the focus moved completely to data practices connected to surveillance. The quadrant graph illustrated both reactive and proactive dispositions, but was mostly centred on (private) data extraction. In this case, the educators moved again from one side to another expressing their diversified positions about data usage (W6, p. 11).

RESEARCHER: *"I think datafication in higher education is..." (open ended statement completed through written post-its)*
DM: *A Bell curve: what really matters (dark for data and eye)*
PD: *Data? Useful*
AK: *A way to decrease education funds and increase the gap between privileged and non-privileged*
WW: *Data for the future: To determine the data present in different activities and use it with educational purposes*
MP: *...is a two-sided blade*
CB: *...a way to kill the art of education*

Most importantly, in this workshop the relevance of technological skills was highlighted as means of preparing for proactive "hacking" of AI biases. While talking about Alexa, one female "tech-savvy" participant said: "...well, actually I have another example because in my house we programmed a minicomputer – with my roommates – to make the voice assistant, but we chose a male voice, and we called it 'Ramón'" (Report W6, p. 13). Indeed, this group discussed the need of a proactive approach in education to develop the skills as "data hackers" in search for public and private data appropriation.

In the Workshop 7 cycle, I kept the perspectives based on public vs private or open vs restricted access and provided the pair of tensions constituted by the reactive (defensive towards data that harms) and the proactive (fostering data practices, even in the case of hacking). This workshop was mainly focused on data in the educational process (particularly learning analytics). The questions raised led us to understand that data practices move from educational research to teaching practice, from the open knowledge as a scholarly endeavour to the protection of students' data, and that more, better and bigger data does not equate to quality research, teaching and learning. Nevertheless, the participants (all coming from the Spanish context) were mostly reliant on the use of data to inform their practices at the beginning of the process (Report W7, p. 10). By the end of the process and during the pandemic, participants were mostly concerned with the failures of the system (unavailable data is a problem, like for example when you use blended approaches, W7). However, against the pandemic situation where all the activities were transferred to the online learning environment, they also concluded that "more data does not equate to more knowledge or an optimal reading of reality", but "good data, shared amongst teachers, could lead to a better educational practice", (W7). This circle contributed to the understanding of educators' professional knowledge on educational data, from the private sphere (students' data care) to the public (as a source of educational reflection).

In the following cycles (from Workshops 8–12), the quadrant graph remained stable, but the participants' engagement helped to saturate or better inform the relevant categories. For example, in Workshop 9, the participants expressed the need to raise (students' and teachers') awareness on data practices based on the familiarisation with technology in educational settings:

> [we as educators should] let them [students] know that everyone is on the same learning curve and that we are all at the same time experiencing something new [data surveillance]. We can improve teaching digital literacies to students if we also improve our understanding of digital literacies.
>
> (AT)

During both Workshops 8 and 10, the need to engage students as data creators and collaborators was also considered. The following conversation expresses this need (W8, p. 14):

ML: Would it be interesting to create student-researcher data communities?

RESEARCHER: You can all create your community for your class and put the data and resources there

AMS: I would love to help!

Above all, at Workshops 11 and 12, I was able to introduce a more complex picture of data practices. I worked together with the participants of Workshop 11 in the elaboration of learning units. Participants integrated technical data skills required to navigate and extract data. They used skills to interpret and critique data visualisations, and they also created data sculptures and data storytelling. In Workshop 12, I provided technical instruments to understand how higher education collect and use students' data, highlighting the limitations of many approaches. I also emphasised the possibility of opening spaces for resistance to detach from dominant discourses about data and build more tailored approaches within the digital classroom. The participants' response again supported the fact that they held different dispositions towards data (seemingly closer to those found in the W7 case). Many participants attended the workshop to collect information on "innovations", others were interested in expanding their technical knowledge. However, there was also a group concerned with data sovereignty, connected to local and regional debates around open-source software in education. This group of participants was interested in curating their data as teachers and researchers, avoiding "academic colonialism" when adopting well-known platforms to teach and to do research:

HPG: in my house I created a Moodle server and Big-Blue-Button where I have all my students' data and it is private.

YMF: What to do when data access is not collectivised and those who have it are not the most suitable to guide an adequate policy?

CPB: What do you think of academic capitalism? Is research (no matter how academic) mediated by economic interests and even personal groups? How to get rid of it?

As a result, the conceptual scheme based on the quadrants was enriched as shown in Figure 5.2

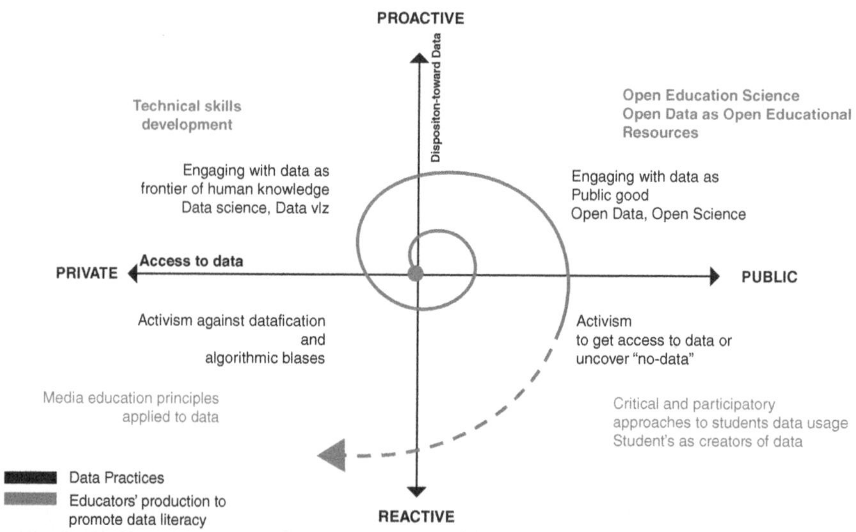

FIGURE 5.2 The results of the iterative process along the workshops through the scheme of the quadrant graph as a complex integrated picture on data practices and literacies.

Discussion: educators' data literacy through a complex lens

In this section, I discuss the evolution and conceptualisation of data practices in education and society, as well as the possible data literacies for educators: see Figure 5.2. This is the main result after the iterative hermeneutic circles of 12 workshops devoted to educators professional learning.

The first quadrant covered the phenomena of open data as a public good and it was based mainly on the results of the Workshops 1 and 4. The reflection arising from it supported the relevance of proactive dispositions towards data as part of the open educational research practice, though some participants were wary about students' data usage. In this case, the required literacies would be browsing open research data, as well as using and generating it. It is worth considering that the literature reports the need to support the open data movement in research as part of Open Science and open access to public knowledge (Borgman et al., 2015). As expressed in the introduction, the proactive dispositions towards data support the idea that data should be used by others, namely, students, scientific communities and citizens (van der Zee and Reich, 2018; Wilkinson et al., 2016). Nonetheless, this is still an issue for most research areas, particularly for education (Raffaghelli and Manca, 2019). Therefore, educational researchers and educators should cultivate this literacy to support fair (open) data practices. However, the participants were concerned about the difficulties of accessing the private platforms to release data to researchers (particularly social media, but also learning management systems). This was deemed an obstacle to furthering open science in education, though it opened up the exploration of a different set of literacies connected with platforms and the unfair use of data.

Moving to the second quadrant (activism to get access to data or uncover no data), the hermeneutic circles (Workshops 2, 3, 9 and 10) focused on open data usage for pedagogical practice. The participants' engagement with the topics of the upper and lower quadrants enriched the idea that open data usage in society, open data in education for a teaching/learning scholarship, and open data as open educational resources require careful collaboration, institutional support, as well as initial training that help educators to understand data quality and the way data might be presented to the students. Nonetheless, in these groups the focus was more on technology use than its critique.

These two circles highlighted the need to promote awareness-raising on reactive dispositions toward data and were concerned about using private data. In the pandemic aftermath, there was an increasing debate around the platformisation and monetisation of education (Williamson et al., 2020). However, by 2019, when the initial bunch of workshops were implemented, this was not *vox populi*. In any case, the Latin American participants expressed a need for participatory approaches and a more democratic usage of data. Indeed, the perspective evolved about how the students might play a role in supporting marginalised populations to collect, interpret and tell their own data stories, even for social problems for which data does not exist. As argued by D'Ignazio and Klein (2020), intersectional data practices face data misrepresentation and overrepresentation of some groups. Using co-liberation, there is a possibility of appropriating advanced data science to support agentic practices. Educators should consider co-liberation when designing for learning with open data, moving from the upper to the lower part of the quadrant's right side.

Moving to the third quadrant (activism against surveillance and algorithmic biases), the Workshop 6 circle clearly detected data surveillance. Indeed, the participants' understandings and practices on data surveillance (both active and passive roles) were evident. I noticed that those with advanced technological skills felt their practices as more "agentic" and empowered to resist, hack or even generate alternative technology development and usage. Aligning with critical pedagogy, technological understanding (or basic literacy) should not be an instrument of adaptation to the system but of resistance and activism (Tygel and Kirsch, 2016). Across the majority of workshops, those with less technical skills pointed out that educators are not "algorithms makers", and they are left with little space for intervening in what they defined as an overwhelming acceptance of datafication. This finding was coherent with Pangrazio and Selwyn (2019). However, as shown in Workshop 7, educators felt confident enough to protect their students' interests, although this action could only cover the educational spaces controlled by the educator, and probably could not be extended to the several spaces connected to students' lives.

In this sense, the educators' professional development also requires data literacy around the surveillance systems, data ethics and alternative software protecting students' data. As it has been pointed out, this relates to the software they work with as educators (Wasson et al., 2016), but also to surveillance, which in society has become a topic that can be developed in class as part of familiarisation with technology (Stewart, 2020) and media education (Hobbs, 2020). During the pandemic, the Workshop 11 circle supported this reflection as well. The participants' attention towards the problem of datafication and data practices was very focused at that point since it was also connected with the so-called "COVID-19 infodemic" and the consequences of ill use of social media.

This circle highlighted the need to enter into the fourth quadrant (the engagement with data as a frontier of knowledge), and it was also supported by the findings in Workshop 12 for educators, in order to embrace a complex approach to data practices, and to intervene accordingly in the development of the data literacies needed, the technical awareness and creativity must be put in tension with the political and ethical concerns. Building on Bhargava and D'Ignazio's concept of popular data literacy (D'Ignazio and Bhargava, 2015), the distribution of symbolic power and material resources through education requires the educators' interaction with institutions to build situated approaches that help understand data and how to use it creatively.

Nevertheless, the quadrant graph would also allow the educator to enter from one or the other side, from the lower or the upper part, covering what is needed at a particular point of the curriculum or from the specific subject field. The quadrant, hence, can be an approach to explore and integrate different forms of data literacy.

Conclusions

In this chapter, I argued that data as social, political, technical and educational problem is complex and evolving. I contend that a situated and explorative approach to understanding data should be considered. This culminated in a conceptual tool built with 292 educators engaged in 12 workshops, to unveil data practices' complexity and facilitate further educators' professional interventions in the promotion of data literacy. This process of understanding the data landscape in search for the bigger picture yielded four types of social activity around data, as experienced by the educators and me: open data as a public good, activism to access existing data or uncover misrepresentation (no data), activism against dark datafication and algorithmic biases, and the engagement with data as a frontier of knowledge. The literacies needed in each of the quadrant's spaces were specific, developed over time and related to the faculty, students and other stakeholders' efforts to make sense of the several dispositions towards data as a base for data politics and practices (Raffaghelli et al., 2020). However, such literacies cannot be seen as separated pieces, but rather, as a complex puzzle.

As one might presume, the institutional scenarios can be various. Discourses on the promising logic of big and open data could be heralded by one institution, whereas in others there could be complete silence and lack of policies around the issue. In the best-case scenario, there might be a participatory approach towards the exploration of the problem to generate guidelines, policies, manifestos or strategic interventions in order to deal with datafication and surveillance, as well as with data for empowerment and co-creation (Raffaghelli, 2020b). Overall, data practices do not relate only to the availability of technological devices or the connectivity across networks, and neither to data processing performance in terms of volume and time. Data practices instead might be connected to participants' dispositions to data.

By exploring the four quadrants proposed here, a group or an institution might be prepared to plan a more strategic professional development for educators. The quadrants could address planning, decision-making, interventions and even activism relating to institutional and individual choices on the digital ecosystem and its consequences. This might also disentangle personal, professional and collective dispositions toward data.

Is data good or bad for our societies, and for educational activity? Like our data practices, this requires a nuanced perspective. This means building integrated pathways towards data literacy as situated learning about data practices. To this end, educators and educational researchers should collaborate with policymakers, activists, the private sector and society, building commitment towards data-driven designs, processes and data products connected to the topics taught and within the educational institution. Concretely, it means to have a deep understanding of what the workflow of data-driven, algorithm-design, coding, data visualisation (proactive data epistemology, both private and public access to data) implies, as well as of the institutional and social impacts of data practices (reactive data epistemologies, both private and public access to data).

In Gadamer's terms, "we learn a techne and can also forget it. But we do not learn a moral knowledge, nor can we forget it" (1960/2013, p. 327). Hermeneutical consciousness (conceived as understanding leading the right action) "is involved neither with technical nor moral knowledge, but these two types of knowledge" (p. 325). Indeed, data practices require a technical, but also a political, aesthetic and ethical dimension, moving from the restricted to the public spaces, and from pro-active to reactive dispositions toward data. Ergo, data literacy can only stem from exploring, interpreting and building understanding around a bigger picture – a complex, integrated perspective on data practices.

Notes

1 Professional learning ecologies for digital scholarship: modernising higher education by supporting professionalism – Ramón y Cajal Programme RYC-2016-19589 (http://edulab.uoc.edu/en/projects/led-projects/professional-learning-ecologies-for-digital-scholarship-modernizing-higher-education-by-supporting-professionalism/).
2 The participants' profiles (after anonymisation procedures) quoted in this paragraph are presented in annex I.
3 In this section, reports cited are listed in the Table 5.1.

References

Baack, S. (2015). Datafication and empowerment: How the open data movement re-articulates notions of democracy, participation, and journalism. *Big Data & Society, 2*(2). doi: 10.1177/2053951715594634

Borgman, C. L., Darch, P. T., Sands, A. E., Pasquetto, I. V., Golshan, M. S., Wallis, J. C., & Traweek, S. (2015). Knowledge infrastructures in science: Data, diversity, and digital libraries. *International Journal on Digital Libraries, 16*(3–4), 207–227. doi: 10.1007/s00799-015-0157-z

Carr, W. (2007). Educational research as a practical science. *International Journal of Research & Method in Education, 30*(3), 271–286. doi: 10.1080/17437270701614774

Carretero, S., Vuorikari, R., & Punie, Y. (2017). The Digital Competence Framework for Citizens With eight proficiency levels and examples of use. *Brussels*. doi: 10.2760/38842

Chi, Y., Jeng, W., Acker, A., & Bowler, L. (2018). Affective, behavioral, and cognitive aspects of teen perspectives on personal data in social media: A model of youth data literacy. In *Lecture Notes in Computer Science (including subseries Lecture Notes in Artificial Intelligence and Lecture Notes in Bioinformatics)* (Vol. 10766 LNCS, pp. 442–452). Springer Verlag. doi: 10.1007/978-3-319-78105-1_49

Corbin, J., & Strauss, A. (2007). *Basics of Qualitative Research: Techniques and Procedures for Developing Grounded Theory* (Vol. 14). Thousand Oaks, CA: SAGE Publications.

D'Ignazio, C., & Bhargava, R. (2015). Approaches to building big data literacy. In *Bloomberg Data for Good Exchange* (online). Boston, Massachussets: MIT Press. Retrieved from https://dam-prod.media.mit.edu/x/2016/10/20/Edu_D'Ignazio_52.pdf

D'Ignazio, C., & Klein, L. F. (2020). *Data Feminism*. Boston, Massachussets: MIT Press. doi: 10.7551/mitpress/11805.001.0001

Eubanks, V. (2018). Algorithms designed to fight poverty can actually make it worse. *Scientific American, 319*(5), 68–71. doi: 10.1038/scientificamerican1118-68

Freire, P. (2000). *Pedagogy of the Oppressed: 30th Anniversary Edition*. Bloomsbury Academic.

Gadamer, H.-G. (2013). *Truth and Method*. London: Bloomsbury.

Gould, R. (2017). Data literacy is statistical literacy. *Statistics Education Research Journal, 16*(1), 22–25. Retrieved from https://iase-web.org/documents/SERJ/SERJ16(1)_Gould.pdf

Gray, J., Gerlitz, C., & Bounegru, L. (2018). Data infrastructure literacy. *Big Data & Society, 5*(2). doi: 10.1177/2053951718786316

Hobbs, R. (2020). Propaganda in an age of algorithmic personalization: Expanding literacy research and practice. *Reading Research Quarterly, 55*(3), 521–533. doi: 10.1002/rrq.301

Lehtiniemi, T., & Ruckenstein, M. (2019). The social imaginaries of data activism. *Big Data & Society, 6*(1). doi: 10.1177/2053951718821146

Lupton, D., & Williamson, B. (2017). The datafied child: The dataveillance of children and implications for their rights. *New Media & Society, 19*(5), 780–794. doi: 10.1177/1461444816686328

Malik, M. M. (2020). A hierarchy of limitations in machine learning. Retrieved from http://arxiv.org/abs/2002.05193

Manca, A., Atenas, J., Ciociola, C., & Nascimbeni, F. (2017). Critical pedagogy and open data for educating towards social cohesion. *Italian Journal of Educational Technology, 25*(1), 111–115. doi: 10.17471/2499-4324/862

Mandinach, E. B., & Gummer, E. S. (2016). What does it mean for teachers to be data literate: Laying out the skills, knowledge, and dispositions. *Teaching and Teacher Education, 60*, 366–376. doi: 10.1016/j.tate.2016.07.011

Markham, A. N. (2018). Critical pedagogy as a response to datafication. *Qualitative Inquiry*. doi: 10.1177/1077800418809470

Maybee, C., & Zilinski, L. (2015). Data informed learning: A next phase data literacy framework for higher education. *Proceedings of the Association for Information Science and Technology, 52*(1), 1–4. doi: 10.1002/pra2.2015.1450520100108

Milan, S., & van der Velden, L. (2016, October 10). *The Alternative Epistemologies of Data Activism*. Retrieved from https://papers.ssrn.com/sol3/papers.cfm?abstract_id=2850470

O'Neil, C. (2016). *Weapons of math destruction: how big data increases inequality and threatens democracy*. New York: Penguin.

Pangrazio, L. (2016). Reconceptualising critical digital literacy. *Discourse, 37*(2), 163–174. doi: 10.1080/01596306.2014.942836

Pangrazio, L., & Selwyn, N. (2019). 'Personal data literacies': A critical literacies approach to enhancing understandings of personal digital data. *New Media and Society, 21*(2), 419–437. doi: 10.1177/1461444818799523

Pangrazio, L., & Selwyn, N. (2020). Towards a school-based 'critical data education'. *Pedagogy, Culture and Society*. doi: 10.1080/14681366.2020.1747527

Papavlasopoulou, S., Giannakos, M. N., & Jaccheri, L. (2017). Empirical studies on the Maker Movement, a promising approach to learning: A literature review. *Entertainment Computing*. doi: 10.1016/j.entcom.2016.09.002

Perrotta, C., Gulson, K. N., Williamson, B., & Witzenberger, K. (2020). Automation, APIs and the distributed labour of platform pedagogies in Google Classroom. *Critical Studies in Education, 00*, 1–17. doi: 10.1080/17508487.2020.1855597

Popat, S., & Starkey, L. (2019). Learning to code or coding to learn? A systematic review. *Computers and Education*. doi: 10.1016/j.compedu.2018.10.005

Pötzsch, H. (2019). Critical digital literacy: Technology in education beyond issues of user competence and labour-market qualifications. *TripleC, 17*(2), 221–240. doi: 10.31269/triplec.v17i2.1093

Prinsloo, P. (2020). Data frontiers and frontiers of power in (higher) education: A view of/from the Global South. *Teaching in Higher Education, 25*(4), 366–383. doi: 10.1080/13562517.2020.1723537

Pybus, J., Coté, M., & Blanke, T. (2015). Hacking the social life of Big Data. *Big Data & Society, 2*(2). doi: 10.1177/2053951715616649

Raffaghelli, J. E. (2018). Educators' Data Literacy Supporting critical perspectives in the context of a "datafied" education. In M. Ranieri, L. Menichetti, & M. Kashny-Borges (Eds.), *Teacher education & training on ict between Europe and Latin America* (pp.91–109). Roma: Aracné. doi: 10.4399/97888255210238

Raffaghelli, J. E. (2020a). Datificación y Educación Superior: Hacia la construcción de un marco para la alfabetización en datos del profesorado universitario. *Revista Interamericana de Investigación, Educación y Pedagogía, RIIEP, 13*(1), 177–205. Retrieved from https://revistas.usantotomas.edu.co/index.php/riiep/article/view/5466

Raffaghelli, J. E. (2020b). Is data literacy a catalyst of social justice? A response from nine data literacy initiatives in higher education. *Education Sciences, 10*(9), 233. doi: 10.3390/educsci10090233

Raffaghelli, J. E., & Manca, S. (2019). Is there a social life in open data? The case of open data practices in educational technology research. *Publications*, 7(1). doi: 10.3390/publications7010009

Raffaghelli, J. E., Manca, S., Stewart, B., Prinsloo, P., & Sangrà, A. (2020). Supporting the development of critical data literacies in higher education: Building blocks for fair data cultures in society. *International Journal of Educational Technologies in Higher Education, 17*(58). doi: 10.1186/s41239-020-00235-w

Raffaghelli, J. E., & Stewart, B. (2020). Centering complexity in 'educators' data literacy' to support future practices in faculty development: A systematic review of the literature. *Teaching in Higher Education, 25*(4), 435–455. doi: 10.1080/13562517.2019.1696301

Ricaurte, P. (2019). Data epistemologies, the coloniality of power, and resistance. *Television & New Media, 20*(4), 350–365. doi: 10.1177/1527476419831640

Sancho-Gil, J. M., Rivera-Vargas, P., & Miño-Puigcercós, R. (2020). Moving beyond the predictable failure of Ed-Tech initiatives. *Learning, Media and Technology, 45*(1), 61–75. doi: 10.1080/17439884.2019.1666873

Stephen-Davidowitz, S. (2017). *Everybody Lies. Big Data, New Data, and What the Internet Can Tell Us About Who We Really Are.* New York: Harper Collins.

Stewart, B. (2020). The open page project. *Journal of Teaching and Learning, 14*(1), 59–70. doi: 10.22329/jtl.v14i1.6265

Tour, E., Creely, E., & Waterhouse, P. (2021). "It's a black hole...": exploring teachers' narratives and practices for digital literacies in the adult EAL context. *Adult Education Quarterly*. doi: 10.1177/0741713621991516

Tygel, A. F., & Kirsch, R. (2016). Contributions of Paulo Freire for a critical data literacy: A popular education approach. *The Journal of Community Informatics, 12*(3). Retrieved from http://www.ci-journal.net/index.php/ciej/article/view/1296

van der Zee, T., & Reich, J. (2018). Open education science. *AERA Open, 4*(3). doi: 10.1177/2332858418787466

Wasson, B., Hansen, C., & Netteland, G. (2016). Data literacy and use for learning when using learning analytics for learners. In S. Bull, B. M. Ginon, J. Kay, M. D. Kickmeier-Rust, & M. D. Johnson (Eds.), *Learning Analytics for Learners, 2016 workshops at LAK* (pp.38–41). Edimburg: CEUR. Retrieved from http://ceur-ws.org/Vol-1596/paper6.pdf

Wilkerson, M. H., & Laina, V. (2017). Youth reasoning with interactive data visualizations. In *Proceedings of the 2017 Conference on Interaction Design and Children – IDC '17* (pp.411–416). New York, USA: ACM Press. doi: 10.1145/3078072.3084302

Wilkinson, M. D., Dumontier, M., Aalbersberg, Ij. J., Appleton, G., Axton, M., Baak, A., ... Mons, B. (2016). The FAIR Guiding Principles for scientific data management and stewardship. *Scientific Data, 3*, 160018. doi: 10.1038/sdata.2016.18

Williamson, B., Eynon, R., & Potter, J. (2020, April 2). Pandemic politics, pedagogies and practices: Digital technologies and distance education during the coronavirus emergency. *Learning, Media and Technology*. doi: 10.1080/17439884.2020.1761641

Willis, J. E., Slade, S., & Prinsloo, P. (2016). Ethical oversight of student data in learning analytics: A typology derived from a cross-continental, cross-institutional perspective. *Educational Technology Research and Development*, *64*(5), 881–901. doi: 10.1007/s11423-016-9463-4

6

DATAFICATION, EDUCATIONAL PLATFORMS AND PROCEDURALISED IDEOLOGIES

Earl Aguilera and Roberto Santiago de Roock

Introduction

As digital technologies become increasingly ubiquitous in everyday institutional participation, so do the associated transformations of everyday experiences into data – often referred to as datafication – including in education. Research on datafication emphasises the ways that human actions and behaviours have been translated into quantifiable metrics, which forms a central role in modern institutions (Foucault, 1977) and increasingly happens through digital technologies (Eubanks, 2018; de Roock, 2020; Gerrie, 2003). However, much research on datafication overlooks the interactional and dynamic nature of the datafied systems and the complex responses of individuals, which are simultaneously agentive and constrained by designed systems embodying the ideologies of their creators (Eubanks, 2018; Noble, 2018; O'Neil, 2016). This chapter unpacks some of these complexities in the context of online learning platforms through the lens of *proceduralisation*. We use this term to describe the ways that ideological constructions (such as social practices, cultural viewpoints, and implicit biases) are embedded in and circulated through the designs of technologies. We then draw on mediated discourse analysis (MDA; Scollon, 2002) in applying our conceptual lens to data examining the online coursework experiences of students at a public university in the western United States who self-identified as the first in their family to attend college (Aguilera, 2019). We analysed interview data constructed around students' experiences in the Canvas LMS, combined with our own walkthroughs (Light et al., 2018) of instructor-facing features of the platform, supplementing these approaches with officially-sanctioned documentation regarding the intended use of various platform features from the official website of the platform developers. We close by discussing pedagogical implications for the material and ideological responses of those immediately involved with datafication through learning systems. We argue that, if the

DOI: 10.4324/9781003136842-6

question is how to live better with data, then educations, institutions, and students must grapple with the ways that data does not just objectively "represent" the realities we experience; instead, our conceptions of data and processes of datafication procedurally and ideologically shape our experience of that reality.

Proceduralising ideologies

To develop a theory of *proceduralisation*, we draw on Bogost's (2010) concept of procedural rhetoric, which builds on the foundations of rhetoric as the art of making (persuasive) claims about the world. Procedural rhetoric extends logocentric and even multimodal traditions of rhetoric to encompass the art of persuasion through processes, such as the rule systems that govern a videogame, a legal proceeding or an economic transaction. While not always computational in nature, these systems also shape our ideological visions of the social, material and virtual worlds that we inhabit. These social realities are not ontological "real" objects waiting to be "discovered". Instead, they are constructed ideologically by groups, communities, institutions or cultures. Thus, digital interactive software, including online LMSs, social media networks and videogames do not solely make claims about the world, but also *construct* ideological visions of the social worlds we inhabit, enrol users into these social worlds, and play a role in maintaining or transforming material relations and power asymmetries.

The development and widespread application of computational technology, which is often referred to as "digital", has inexorably transformed the nature of proceduralisation by developing computational media forms that are at once encyclopaedic in scope, spatially navigable, and inherently participatory, in addition to being process-based in nature (Murray, 1997). Drawing on Murray's four-part characterisation of computational media, we understand proceduralisation as operating in at least four ways:

- **Repetition**. Unlike a book, which you only read once or a few times, regularly used software such as a social media platform or LMS do ideological work through repeated everyday use. The more often used, the more potential for circulating and shaping ideologies.
- **Scale**. Datafication, often drawing on large-scale data sets, relies on software that extracts information from many, many people. Similarly, the more people are exposed to a piece of software that does ideological work, the stronger the potential for influence.
- **Automaticity**. Computational software relies on processes that are triggered automatically given certain human inputs, giving the feeling of "naturalness" when using digital technologies.
- **Invisibility**. Unlike multimodal representation, computational processes underlying software are usually imperceptible by humans. Rendering these processes invisible makes it difficult to understand, let alone resist or critique their influence.

While there are many things that can be proceduralised through computational means, we focus in this chapter on the ways that ideological constructions, such as social practices, cultural views, and implicit biases are embedded in, and enacted through educational software.

We follow cultural theorist Stuart Hall in understanding ideology as "mental frameworks – the languages, the concepts, categories, imagery of thought and the systems of representation – which different classes and social groups deploy in order to make sense of, define, figure out and render intelligible the way society works" (Hall, 1986). Van Dijk simplifies this definition even further, defining ideology as "the basis of social representations shared by members of a social group" (1998, p.8). Like Hall et al., we are concerned with the connection between ideology, power and hegemony, namely, the ways particular sets of ideologies come to dominate the thinking in a particular historical period that aid in the maintenance of power and material relations, in this case through notions of teaching and learning (Abercrombie and Turner, 1978).

Bringing these ideas together, we offer the concept of *proceduralised ideologies* as a way to describe the particular ideological constructions that are embedded in and propagated through procedural means within technological designs. Examples of proceduralised ideologies include the settler-colonial cities (but not indigenous territories) that are recognised as "valid" locations in mapping software, the range of emotional ideographic responses (i.e., emojis) that are included or excluded in videoconferencing software, and the range of linguistic accents that are considered intelligible within speech-to-text software. With regard to teaching and learning, these proceduralised ideologies can include the roles that people adopt for themselves and position other people in society, including what it means to be a teacher or a student. When these proceduralised ideologies are not critically examined, we run the risk of allowing the worldviews, political economies, and power relations that characterise the experiences of software developers to frame more and more of our technologically-mediated social practices in ways that maintain or exacerbate existing inequalities (Benjamin, 2019; de Roock, 2020).

While the sudden jump to widespread online learning precipitated by the COVID-19 pandemic has raised the stakes of this phenomenon within education, we instead see this pandemic as accelerating historical trends of ideological proceduralisation in education. And while the phenomenon is often lumped into a general "online learning" trope, we are particularly concerned about how the use of educational software is shaping contemporary views of what counts as teaching and learning.

Methodology

Approaching through the lens of proceduralised ideologies, we now turn to an examination of learning management systems (LMsSs) as sites through which the fundamentally human endeavours of teaching and learning are standardised, datafied and commodified. We approached this analysis with the following question in mind: *How are the social actions of "teaching" and "learning" ideologically constructed and procedurally enacted through the design and use of the learning management system?* To address this questions, we undertook a mediated discourse analysis (MDA; Scollon, 2002) of interview data constructed around students' experiences in the Canvas LMS, combined with our own walkthroughs (Light et al., 2018; de Roock and Baildon, 2019) of instructor-facing features of the platform, supplementing these approaches with officially-sanctioned documentation regarding the intended use of various platform features from the official website of the platform developers.

Context and Participants

The analysis developed in this chapter primarily involves interview data collected from a broader study focused on understanding the online coursework experiences of students who self-identified as the first in their family to attend college (Aguilera, 2019). As that project demonstrated, students identified as first-generation for a variety of reasons, some of which intersected with issues of gender identity, racialisation and citizenship status. The initial sample of interviews included eight students, including one graduate student. The sample aimed for diversity in gendered, racialised, and academic/disciplinary experiences, as well as prior experiences with online courses. Participants were all attendees of a public university in the Western United States, which is part of a statewide comprehensive public university system. The university is designated as both a Hispanic Serving Institution (HSI) and an Asian-American Native American Pacific Islander Serving Institution (AANAPISI), with about 3 in 4 of incoming students in the Fall 2020 semester identifying as African American, Native American, Asian, Pacific Islander or Hispanic, along with a high percentage of (self-identified) first-generation Pell Grant students, often equated with being low-income.

Through our initial recruitment surveys and the interviews with students, we learned of the diverse educational, cultural and social experiences our participants brought with them into their online coursework. A summary of participant demographics, primarily drawn from their responses to the recruitment survey, is shown in Table 6.1.

Data Sources

The initial construction of the dataset involved conducting three interviews per participant, resulting in a total of 24 sets of interview data. Over three rounds of interviewing, participants discussed their prior and current experiences of their online courses. The second of the three interview rounds focused explicitly on their current online course experiences, and included a screen-recorded walkthrough (Light et al. 2018) and discussion of the ways they accessed and navigated the LMS in their role as students. All interviews were recorded and transcribed. Both transcribed interviews and audio/video recordings were used as artefacts of the current analysis. For the purposes of this analysis, we focus on illustrative examples

TABLE 6.1 Participant demographic summary from recruitment survey

Participant Pseudonym	*Degree Level*	*Major/Programme of Study*	*Year in Programme*	*Year of Birth*	*Gender*	*Race/Ethnicity*	*Prior Online Course Experience?*
Montana Rufin	Undergraduate	Political science and Sociology	Fifth Year or More	1997	Female	Latinx	Yes
Indigo Serrano	Undergraduate	Liberal Studies	Fourth Year	1996	Female	Latinx	Yes
Eve Phang	Undergraduate	Psychology	First Year	2000	Female	Asian	No
Esmeralda Ortega	Undergraduate	Biochemistry	Second Year	2000	Female	Latinx	No
Osvaldo Iglesias	Undergraduate	Sociology	Second Year	1999	Male	Latinx	Yes
Lin Chiu	Graduate	Master of Science in Counselling Option in Student Affairs and College Counselling	Fourth Semester or More	1989	Female	Asian	Yes
Ayesha Singh	Undergraduate	Kinesiology, Exercise Science	First Year	2001	Female	Asian	Yes
Adriana Castro	Undergraduate	Liberal Studies	Third Year	1985	Female	Latinx	Yes

of patterns we identified across multiple interviews, though the individuality and idiosyncrasy of each student experience should be noted.

Mediated Discourse Analysis

Our methodological lens draws primarily from MDA (Scollon, 2002). MDA, also referred to as nexus analysis, has been described as an approach to linguistic discourse analysis that emphasises the relationship between discourse and action with a particular emphasis on underlying ideologies. In comparison with other approaches to discourse studies more broadly, MDA emphasises social action as an object of focus, framing linguistic discourse as just one mediational means through which social actions can occur (S. Scollon and de Saint-Georges, 2012). MDA and related approaches have been used to analyse indigenous language distance education and activist identities (Scollon and Scollon, 2004), the ideological "calling out" of identities among African American Twitter users (Brock, 2018), the recruitment of particular identities by smartphone dating apps (Light, Burgess, and Duguay, 2018), and the neoliberal ideologies of STEM (science, technology, engineering, and mathematics) imposed on young students through a state-sponsored online learning portal (de Roock and Baildon, 2019).

We draw on MDA as a means for analysing the ways in which multimodal discourse involving those positioned as teachers and those positioned as students, along with procedural design (e.g., the kinds of actions the LMS-as-designed-artefact affords and constrains) come together to produce a particular ideological enactment of the social actions of teaching and learning. Keeping these definitions and philosophical commitments in mind, we next highlight how we engaged in nexus analysis, the primary methodological approach of MDA.

Engaging and navigating the nexus of practice

The opening task of any nexus analysis, *engaging the nexus of practice*, consists of the researchers positioning themselves as part of the nexus of practice under study in what is sometimes called a "zone of identification". For our project, delineating this zone of identification first involved identifying the particular nexus of practice that would be the focus of our analysis. As our research questions allude to, we chose to focus on the intertwined practices of teaching and learning in online courses, specifically in the design and actual use of the Canvas Learning Management System within our research context. To further, map a zone of identification, we examined recruitment surveys for general demographic information. We also reviewed initial interviews to develop a better sense of the participant's educational experiences within and beyond online learning. Taken together, these prior accumulated experiences are referred to in MDA as the "historical body". Finally, part of our engagement with the nexus of practice was surveying our access to the features of the LMS software that would give us a closer understanding of the different experiences students would be describing in their interviews. This primarily involved engaging with the teacher-accessible features of the software platform itself, as well as engaging with administrator-facing tutorial documentation from Instructure, though neither of us had access to administrator-level features.

The second main phase of our nexus analysis, *navigating the nexus of practice*, focused mainly on the second round of interviews conducted with our eight participants. Items in these interviews focused on participants' specific uses of the Canvas LMS in their online coursework, which gave us a glimpse into how their instructors were enacting particular ideological models of teaching through the platform. These second-round interviews included video-recorded "screen-sharing" by participants, in particular when we asked them to walk us through a typical session in their online courses – what they did from sign-on (generally a first step to accessing the LMS) to sign-off (indicating they were done engaging with the coursework for the day). In our analysis, we worked to try and disentangle the specific practices (e.g., completing quizzes, posting to discussion boards, sending private questions to instructors) that constituted the students' engagement with a particular ideological vision of learning. This disentangling of practices within the nexus is referred to in MDA as the "interaction order".

Following the interaction order identified by students in these interviews, we then set to the task of navigating what these different features (quizzes, discussion boards, private messaging) looked like from the perspective of those with teacher roles/privileges inside the LMS – features that the two of us had access to as professors at our respective institutions, but which students did not have access to. We consulted tutorial documentation provided by Instructure for teachers, students, and even administrators (though we did not have direct access to administrator features in the platform) to get a sense of the intended use of these different features – which we argue offers glimpses into the ideological visions of teaching and learning that have been proceduralised into the LMS.

We treated these three sources of data: student interviews, software walkthroughs, and software documentation as sites for examining the ways that language was being used as this intersection of social action (in our case, online teaching and learning). This collection of *language-in-use-at-the-intersection-of-social-action* is referred to in MDA as "discourses-in-place", and served as a key focus of our analysis. Throughout all of this phase, we documented initial hypotheses, composed research memos, discussed and examined these hypotheses in light of confirming and disconfirming evidence, and refined our hypotheses into the claims we will present as the findings of this project (Erickson, 1986).

Analysis and Findings

In the section that follows, we organise findings of this analysis around the key conceptual areas at the theoretical core of MDA: historical bodies, interaction order, discourses-in-place and social action. With regard to *historical bodies* in the nexus of practice, we found that contrary to discourses about the "disruptive" potential of educational technology, the socially enacted roles of "student" and "teacher", were further reified through the procedural ideologies and use of the Canvas LMS for facilitating online coursework. With regard to *interaction order,* we found that the Canvas LMS was designed with an ideological orientation towards a highly teacher-controlled approach to instructional design, through which students engaged in practices that could "game the system" to serve their own goals and enact alternative ideological visions of achievement. Finally, with regard to *discourses-in-place*, we found that the Canvas LMS often adopted and proceduralised dominant ideological discourses of institutional schooling, while participants enacted their role as students in the course by performing discourses of pedagogical engagement.

Historical bodies: Proceduralising ideological conceptions of "student" and "teacher"

Our analysis of historical bodies within this particular nexus of practice considered both *people who were positioned as students*, and *people who were positioned as teachers* in the ideological enactment of online coursework. We found that positioning was facilitated through at least two sets of processes:

1) The existing ways these roles were socially determined and enacted outside of online platforms (i.e., the traditional "grammars" of schooling).
2) The ways in which these roles are procedurally determined and enacted through the design and implementation of the Canvas LMS.

In our analysis of student interviews, we found the existing institutionally determined roles of teachers and students evidenced primarily through the practices students appeared to associate with their instructors, and those they associated with themselves as participants in the course. Montana, a fifth-year political science and sociology major, for example alludes to some of these practices in the following passages:

STANZA 1

1. So //
2. this is like the main page /
3. Like where our core summary //
4. and...
5. these are all the announcements
6. So as you can see like //
7. She was like /
8. a little off track a bit /
9. But then she updated the entire class //
10. and...
11. these are all the assignments we have to turn in so //
12. we have like four /
13. Four assignments that are coming up within //
14. for the rest of the semester...

Montana's use of "she" in Stanza 1, lines 7 and 9 refer to the instructor for the course. In lines 7–8, Montana associates the instructor with the action of updating the entire class with "assignments we have to turn in" (Line 11) – the "we" here referring to Montana and her peers. The notion that "She was a little off track a bit" (Line 8) seems to refer to the instructor not fulfilling an expectation of keeping the course's assignments posted and organised. On the other hand, Monica associates her role and that of her peers with the actions of turning in assignments, noting in particular the load of work outlined by the instructor for the students to complete for the rest of the semester (Lines 13–14).

Montana further discusses other actions that constitute the role of "student" in this space in the passage that follows:

STANZA 2

1. And then... this is a discussion board /
2. So like this is where we would have like our discussions that we have to post /
3. At least once a week /
4. But now we only shifted to coronavirus //

Here, Montana highlights the expectations for the usage of the "discussion board" feature of the Canvas LMS featured in her online course. As part of the practice of being a student, at least in Montana's class, is to post on the discussion board at certain time intervals (Stanza 2, Line 3). However, the Killer Microbes course that Montana is enrolled in shifted to focus on a single discussion board (Stanza 2, Line 4), which she details here:

STANZA 3

1. This one's like a //
2. Long-term discussion board //
3. So you have to post at least once //
4. and it has to be a meaningful post //
5. So like my other classmates already posted stuff /
6. And then /
7. You're supposed to respond to someone else's and it's worth 10 points...

In construing her role as student fulfilling the expectations of the instructor, Montana highlights the acts of posting to the discussion board, along with a reference to frequency (Stanza 3, Line 3) and some measure of quality (Stanza 3, Line 5). Further, as she discusses the expected procedure of participation in Stanza 3, Line 7, she also refers to a measurement of student efforts via a point system. Intertwined here is the system of computational procedures that give rise to the "Discussion Board" feature of Canvas as it is designed into the course, as well as an ideological construction of student-ing as a procedural phenomenon, that might be summarised here through the following formulation:

make note of assignments	*complete assignments following instructions*	*receive points*

Across multiple interviews, participants identified a similar set of processes for enacting the role of "student" and "teacher" in their online courses. While these findings alone are not novel – for example, Tyack and Cuban (1995) have identified similar processes as part of the dominant "grammar" of schooling – what we also notice is how these roles are procedurally designed into the Canvas LMS itself through its "roles" function. *Roles,* within the context of software development, refer to sets of permissions given to different users that delineate what a user can/cannot do or access within that software platform. In the Canvas LMS, the Teacher role is described in tutorial documentation as a Course-Level Role, "with permissions that allow a user course-level access. Usually users with these roles cannot see more than what is in their Canvas courses" (Instructure, 2021). Going further in this documentation, the Teacher role is described in this way:

> The Teacher role grants a user course admin permissions, giving them control over their assigned course or courses. However, institutions can revise and limit these permissions as needed, depending upon the needs of your institution.
>
> (Instructure, 2021)

In simple terms, the Teacher role is permitted to access and create course content, including announcements, assignments, discussions, pages and quizzes, as well as the ability to group these elements into modules. The Teacher role is permitted access to grades and other identifying information as well. As the tutorial documentation alludes to in its reference to "permissions", and institutions who have the ability to "revise and limit these permissions as needed", hierarchical relationships are designed into the procedural architecture of the LMS, with the role of *Account Admin* having the highest authority, and students having the least:

> Account admins have the power to set permissions for all users in Canvas. Admins can create additional account-level roles with account-level permissions. The default permissions for account admins can include access to everything within the account, plus the ability to masquerade as a user.
>
> (Instructure, 2021)

We can contrast this with the way the *student* role is described in the same documentation:

> These users are permitted to submit assignments. This permission should not be turned off for this role. Student permissions are restricted, but they have enough permissions to access and interact with course materials. You may also grant other permissions to students.
>
> (Instructure, 2021)

In this passage, taken from an administration-level tutorial, the "you" likely refers to those with an administrative role – typically those representing the interests of the "institutions" mentioned in the teacher role description. The Instructure documentation also provides guidelines for what administrators should (or should not) do, given their administrative privileges. In contrast to discourses circulated about innovative technologies disrupting education, or in Instructure's own promotional materials, "reimagining the classroom", documentation already begins to hint at the ways in which the platform procedurally reproduces existing ideologies of hierarchical institutionalised education.

In the next section of the analysis, we highlight how the procedural design and social enactment of the roles of teacher and student also informed the ways participants interacted with content in their courses.

Interaction order: Constrained agency within a procedural system

The term interaction order, within MDA, refers to a disentanglement of the various actions involved within a particular nexus of practice. Participants in our study identified a variety of practices they engaged in while enacting their role as students, such as completing quizzes, posting to discussion boards, sending private questions to instructors, and submitting assignments. However, students also demonstrated their own "reading" of the procedural design of their courses and highlighted their own ability to "game the system" based on goals they adopted (e.g., maximising time efficiency or increasing their grade average). We can see an instance of this in the ways that Eve Phang, a first-year undergraduate psychology student, engaged with Canvas-integrated textbook-and-quiz software known as InQuizitive. In Eve's second interview, she describes her understanding of how InQuizitive functions within Canvas:

> So the reason why we have to access it through Canvas is because it says, "Connected to LMS", and so that just makes sure that everything you do here is transferred onto your gradebook on Canvas. And so the amount of time that you spend on each chapter is also recorded. So it lets the professor know if you've studied or if you didn't study. Things like that.

In this passage, Eve highlights the ways that her actions as a student – in this case reading the course textbook, are surveilled (through computational tools), datafied (in the form of time spent within each chapter), and then procedurally connected to her evaluation as a student (via the Canvas gradebook), each of which have implications for her experience of what it means to take on the ideologically constructed role of "student".

After further elaboration of the structure of the textbook Canvas connection, Eve highlights the ways she can interact with the built-in quiz software, which consists of multiple-choice questions provided by the textbook publisher, illustrated in Figure 6.1.

> So, how it works is like this. Where it says, "Activity score", that basically… it shows how many times you got it wrong, how many times you got it right. And it kind of shows your progress. And then this is the confidence interval I was talking about. So you're kind of betting your knowledge seeing how well you know. So let's say I know it all the way or at least I say I do, and I click this and it's oh, that's right. Okay. I didn't know that was right. Okay. But if it's right, then you gain 100. If it's wrong, then you lose 100. But if you eventually get it right, you get 50 points for it. It's like that. But let's say like I get it wrong. Oh, okay. Never mind. Wrong. Then you lose 100.

Considering Eve's outline, we can see several designed features of this quiz software coordinating to ideologically construct the experience of being a student in the online course. First, there is a scoring system that takes into account "how many times you got it wrong, how many times you got it right". Then, there is a kind of wagering system ("the confidence interval") that encourages students to influence their potential score based on their confidence ("But if it's right, then you gain 100.

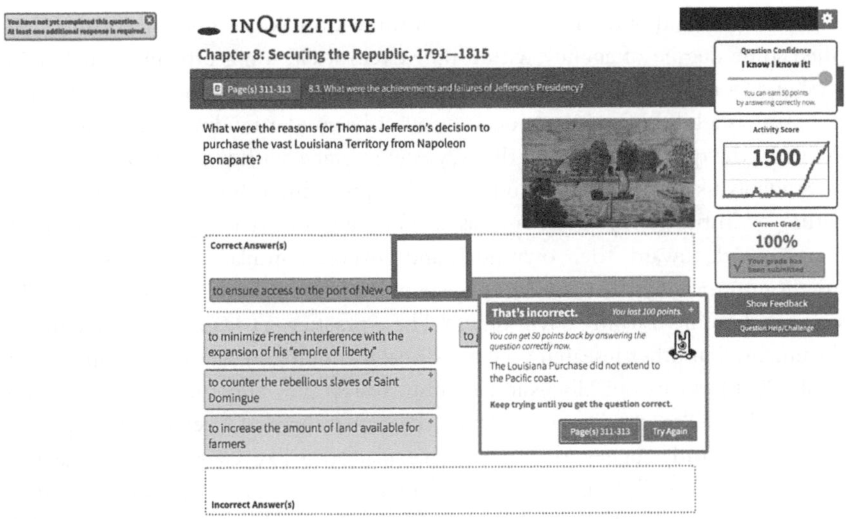

FIGURE 6.1 An image of Eve's in-course quiz software view.

If it's wrong, then you lose 100"). The questions themselves are presented in "classic" multiple choice style, with incorrect answers leading to lost points and corrective visual feedback, as illustrated in Figure 6.1. Finally, an overall grade calculation shows how the particular scoring systems are impacting an overall evaluation of student progress.

Moving on, Eve discusses the ways she is able to manipulate these procedural designs to enact particular goals she can adopt for herself:

> But I mean, since I already got up to the amount of points I needed, my grade stays at 100 no matter what. But if I wanted to go back and review, then I would do so. But like I said, since I got it wrong, it says, "You can get 50 points back by answering the question right", and then it lets you know that the answer I chose wasn't correct and it tells you what pages to go back to, to review, to know the answers and things like that. And so that's pretty much how it works. So, just let me keep going. Okay… here. I think that's wrong. Okay. So next question. And then let's say I want to bring this down lower, the most I can lose, I believe, is up to 20 and the most I can gain is 20 points too. That's how it works.

Note Eve's focus on points and scoring, signalled by utterances like "my grade stays at 100 no matter what", "You can get 50 points back by answering the question right", and "the most I can lose, I believe, is up to 20 and the most I can gain is 20 points too". While she does emphasise learning practices she can engage in ("If I wanted to go back and review"), much more of her response focused on the procedural design of the online quiz platform. Her use of conditional statements such as "if I wanted to", and "let's say" suggest an orientation towards experimenting with the inputs and outcomes of the platform, in a similar way to the way one might explore inputs and outcomes in a game or simulation. We also noted that she identified some degree of agency within the system itself, using utterances such as "it says" and "it lets you know". Near the end of this selection, she indicates "that's how it works" as a kind of summary declaration of her understanding of the procedural design of the quiz platform and the ways she interacts with it as a student.

Students like Eve come to understand the procedural (but non-computational) nature of institutional, classroom-based schooling and develop tactics and practices to work towards their own goals and purposes. Similarly, Eve's case provides insight into the ways that students engage in these practices within a procedurally-structured online learning environment. These practices of "gaming the system" are documented in other research on educational technologies (see, for example, Baker et al., 2008). As in our discussion of historical bodies and the proceduralisation of the roles of "teacher" and "student", the experiences of Eve and other participants of the study suggest that, in practice, platforms like Canvas and InQuizitive seem to procedurally reproduce and ideologically normalise transactional models of schooling which emphasise scoring, metrics, evaluation, and other sorting and classification practices further bolstered by neoliberal institutional discourses of "data-driven instruction". While the experiences of other participants, such as Master's in Counselling student Lin, suggested more open, free-flowing and relational educational experiences, most of these seemed to occur beyond the bounds of the Canvas LMS – in informal discussions, email exchanges and videoconferencing sessions with instructors and peers.

We now turn to an analysis of some of the ways in which participants discursively enacted their roles as students in the online courses, specifically through the MDA concept, discourses-in-place.

Discourses-in-place: Performing student "engagement"

From the perspective of MDA and related traditions of discourse analysis, language does not exist to solely convey information. Instead, it is used, in coordination with multiple communicational forms, as a means for taking action in the world (Rogers, 2011; Scollon, 2002). In the case of this project, we found that one recurring way that participants utilised language in their online coursework was in performing the *discourse of student engagement*. In using this term, we refer once again to widely accepted "grammars of schooling", which include certain ways of doing, being, acting and speaking to perform various roles in institutional educational contexts.

For teachers, this can include the common initiation-response-evaluation pattern that has been documented in studies of classroom discourse (Neal, 2008). For students, such as the participants of our study, this included the performance of "engagement" – an oft-cited but contested construct within educational research (Axelson and Flick, 2010). Building on Axelson and Flick's (2010) simplified definition of "student involvement in a learning process", we noted the ways in which students recognised their performance of this construct through discourse. An exchange between Earl, our first author and Indigo, a fourth-year liberal studies undergraduate student, illustrates this with regard to the Discussion Board function, which has is a commonplace design within LMS systems like Canvas:

EA: Okay. Are there any other ways you interact with your classmates in the course?

INDIGO: No. Are you talking about discussion boards maybe? Or is?

EA: If that feels like an interaction to you? Then, yes.

INDIGO: I mean, it's a fake interaction, fake and forced interaction, but every now and then my Nat Sci will have them, but mostly it's my CRIM class that uses discussion posts.

EA: Got it. Can you tell me a little bit more about the fake and forced nature of that interaction?

INDIGO: I mean, I don't think or know that any of us are genuinely curious. It's strictly business. None of us are trying to make friends on the discussion post. So our teachers telling us to do it and you just got to do it.

Initially in the exchange, Indigo questions the very nature of the practice of Discussion Board-ing as a means of interaction with her peers. When Earl confirms that yes, the discussion board could be addressed, Indigo offers the words "fake" and "forced" to qualify the concept of "interaction" (stating the word "fake" twice) on these boards. Earl inquires further about the terms offered, and Indigo clarifies her position that "I don't think or know that any of us are genuinely curious". Further, her indication that "It's strictly business", reminded us again of the transnational model of online education we discussed in our analysis of interaction order. She alludes to a kind of dispassionate, disconnected experience of learning when mentioning "None of us are trying to make friends on the discussion post". Her remark that "our teachers telling us to do it and you just got to do it" reinforces the proceduralisation of the roles of teacher and student we discussed in our analysis of historical bodies.

A few moments later, Indigo directly references the performative nature of student discourse on the discussion boards:

INDIGO: You got to pretend so that you can get credit.

EA: And okay. And what are you pretending? Or what's the goal of?

INDIGO: I mean, pretending to be engaged. To be like, wow, I love your point of view when it comes to the color of grass. I also agree that the grass is green. Good job. When you wouldn't say that in real life. You're just like, cool.

In addition to highlighting that she (and in her estimations, other students), are "pretending to be engaged", she illustrates her claim with examples, which we interpreted to be an ironic, if not exaggerated framing – minimising the significance of the social practice: "To be like, wow, I love your point of view when it comes to the color of grass. I also agree that the grass is green. Good job." If we interpret this discussion of the colour of grass as an unremarkable fact, then we can interpret the response to such a presentation ("I love your point of view…", "I also agree…", "Good job…") as a disproportionately overenthusiastic response to the mentioning of an unremarkable fact. This, for Indigo, seems to be an illustration of what it means to discursively perform engagement as a student. Indigo contrasts her claim that "you wouldn't say that in real life", with the more concise, but perhaps proportionately enthusiastic utterance: "cool".

We noted these sentiments from other participants as well, such as this response from Esmerelda Ortega, second year biochemistry undergraduate student, after Earl asked about whether there were other ways that she interacted with peers in her online courses:

ESMERELDA: Not usually, but there has been one, well, actually we respond to each other on our discussions. We have to respond to two other people, and I think it's more of like I don't know if platonic's the word. But you don't really get to know them. You get to know their ideas and you're not really interacting with them. It's just like you're responding to what they said. And either if you agree or what you liked about it or if you can suggest something usually like that.

Eve Phang, who we discussed earlier in the chapter, had this to say on the topic of engagement on discussion boards:

> Since it's like every other week and that's every other week regarding the discussion board, I would say never. Even commenting on their thing. I feel like that's not even really engagement. That's just doing something that you're required to do. Because if you don't do it, then you don't get the full points because you're required to answer at least two of your classmates' discussion response and things like that. But I'm totally okay with it, though. Yeah. Yeah.

Finally, first-year kinesiology undergraduate Ayesha Singh contrasted Discussion Board interactions with those that happened in the physical classroom.

AYESHA: So, I feel like the discussion, how we, if we were to have this discussion with someone face to face, we would know how they actually feel about the discussion that they're having and the ones that are typed up like this. We don't know how many times they erase the word before writing it. We don't know how many times they rewrote that sentence, but if it was face to face, they would just say it. It would be natural.

Across these responses, we noted our participants' descriptions of engagement with peers ("you don't really get to know "them"), the compulsory nature of discussion board interactions ("That's just doing something that you're required to do", "We have to respond...") and the differences between the ideologically and procedurally constructed world of the "online course" contrasted with other contexts ("...you're not really interacting with them", "I feel like that's not even really engagement", "if we were to have this discussion with someone face to face..."

While this discourse of performative engagement might be documented in other contexts, we found that in combination with the proceduralisation of roles, the transactional educational models facilitated by the platform, and sometimes socially disconnected nature of online learning, such discursive performativity was characteristic of many of our participants' experiences of online coursework on the Canvas LMS.

Social action: Proceduralised ideologies of "teaching" and "learning" in an LMS

Looking across the questions, we have attempted to highlight that the core social actions we are aiming to understand are *teaching* and *learning*, specifically as they are designed and enacted in an online learning context. Looking across our analysis of historical bodies, interaction order and discourses in place, we summarise the ideologies of teaching and learning within the Canvas LMS with a series of propositions.

> To be a teacher within the LMS is to take on a hierarchical role of power and agency above that of a student, maintained through both historically situated

> social practices, institutional policies and designed computational procedures. This teaching role is procedurally characterised by the development of a course structure, the disseminating of assignments and the evaluation of student progress. It entails the navigation of a transactional model of education facilitated by an online learning platform, with the ability to automate procedures of assessment and collect certain kinds of data about student "progress" as determined by the platform's capacities. Conversely, to be a student within the LMS is to occupy a lower position of power and agency within an institutional and computational hierarchy. This role is characterised by the navigation of course expectations, the completion of assignments and the discursive performance of engagement. It entails the participation in a transactional model of education facilitated by an online learning platform, with the ability to exercise a limited degree of agency to manipulate the system's computational systems of measurement, grading and evaluation via a kind of procedural literacy.

This ideological and proceduralised model of teaching and learning is facilitated by certain designed features of the Canvas LMS (roles, surveillance capacity, datafication) as well as the social practices mediated through the platform (instructional design, teacher-student interactions and student-student interactions). In the final sections of this chapter, we offer some discussion of how these findings might be considered within the broader context of an increasingly datafied educational landscape.

Discussion

The findings of the project are consistent with existing research in online teaching and learning which identifies some of the challenges with socially-disconnected, overly structured, and dehumanising experiences (Kilgore, 2016; Song et al., 2004; Weiss, 2000). We extend this literature by further clarifying the challenges associated with the ways that the automation of these practices, via various procedural designs and uses of LMS platforms, can begin to shape ideologies of concepts and practices like teaching and learning. We have used the term *proceduralised ideologies* to describe this phenomenon, highlighting both the ideologically constructed nature of concepts like "teaching" and "learning", as well as the ways that these ideological constructs are proceduralised through computational processes, software designs and social practices.

While advocates of critical, transformative and imaginative pedagogies have long envisioned more humanistic, politically engaged, and creative ways of "doing school", and while advocates of educational technologies have often touted the "disruptive" potential of digital platforms, our analysis suggests that it is current design and application, online coursework delivered through the Canvas platform does not represent either of these visions. Instead, what stands out to us is the computational proceduralisation and automation of these dominant social actions – a digitised version of the "factory model" of schooling. Of course, we understand that a technological platform on its own does not define an educational experience; however, our analysis also suggests that the ideologies and social practices associated with this particular experience of online teaching and learning have not significantly differed from the historical grammars of schooling that have come to characterise institutional education.

If the question that animates this edited volume is how to live better with data, then our findings suggest that educators and institutions may need to rethink what we consider to be educational data, how this data is generated and collected, what value we place on it, and what role our datafication practices have in shaping our shared understanding of what it means to be educated, to become educated and to educate others. As we have argued, these practices and processes do not just describe the world around us – they can actively shape our shared ideologies, experience and the actions we take in the social, material and virtual worlds we share. Though not all of us may have the technical knowledge of how software like the Canvas LMS is built, we can still understand some of the outcomes – including the proceduralised ideologies, designed into and enacted through this software. In a time when educational processes are becoming increasingly automated, scaled, repeated and rendered invisible by the computational tools adopted to facilitate teaching and learning, we believe that understanding and responding to these proceduralised ideologies will only become more important over time.

References

Aguilera, E. (December 2019). First-Gen College Students' Experiences and Enactments of Digital Literacies in Online Courses. *Paper presented as part of the Scholars of Color Transitioning Into Academic Research Showcase at the Literacy Research Association 2019 Annual Conference.* Tampa, FL.

Aguilera, E. & De Roock, R. (December 2019). De-Centering the Digital: Toward a Politics of Hybridity in Digital-Age Literacy Practices. *Paper presented as part of the symposium, De-Assembling the "Digital": Possible Futures for Digital Literacy Research and Pedagogy at the Literacy Research Association 2019 Annual Conference.* Tampa, FL.

Abercrombie, N., & Turner, B. S. (1978). The dominant ideology thesis. *The British Journal of Sociology, 29*(2), 149–170.

Axelson, R. D., & Flick, A. (2010). Defining student engagement. *Change: The Magazine of Higher Learning, 43*(1), 38–43.

Baker, R., Walonoski, J., Heffernan, N., Roll, I., Corbett, A., & Koedinger, K. (2008). Why students engage in "gaming the system" behavior in interactive learning environments. *Journal of Interactive Learning Research, 19*(2), 185–224.

Benjamin, R. (2019). *Race after technology: Abolitionist tools for the new Jim code.* Social Forces. Meford: Polity Press

Bogost, I. (2010). *Persuasive games: The expressive power of videogames.* MIT Press.

Brock, A. (2018). Critical technocultural discourse analysis. *New Media & Society, 20*(3), 1012–1030.

Cresswell, J. W. (1998). *Qualitative inquiry and research design: Choosing among five traditions.* Thousand Oaks, CA: Sage.

Ekbia, H. R., & Nardi, B. A. (2017). *Heteromation, and other stories of computing and capitalism*. MIT Press.

Erickson, F. (1986). Qualitative methods in research on teaching. In M. C. Wittrock (Ed.), *Handbook of research on teaching* (pp.119–161). MacMillian.

Eubanks, V. (2018). *Automating inequality: How high-tech tools profile, police, and punish the poor*. St. Martin's Publishing Group.

Foucault, M. (1977). *Discipline and punish: The birth of the prison*. Vintage Books.

Garcia, A., & Nichols, T. P. (2021). Digital platforms aren't mere tools – they're complex environments. *Phi Delta Kappan, 102*(6), 14–19.

Gee, J. (2015). *Social linguistics and literacies: Ideology in discourses*. Routledge

Gerrie, J. (2003). Was Foucault a philosopher of technology? *Techné: Research in Philosophy and Technology*, 7(2), 66–73.

Hall, S. (1986). The problem of ideology – Marxism without guarantees. *The Journal of Communication Inquiry, 10*(2), 28–44.

Instructure, Inc. (2020, July 20). *What user roles and permissions are available in Canvas?* Instructure Community. https://community.canvaslms.com/t5/Admin-Guide/What-user-roles-and-permissions-are-available-in-Canvas/ta-p/102

Kilgore, W. (2016). *Humanizing online teaching and learning*. Pressbooks.

Light, B., Burgess, J., & Duguay, S. (2018). The walkthrough method: An approach to the study of apps. *New Media & Society, 20*(3), 881–900.

Murray, J. H. (1997). *Hamlet on the Holodeck: The future of narrative in cyberspace*. MIT Press.

Neal, M. (2008). Look who's talking: Discourse analysis, discussion, and initiation-response-evaluation patterns in the college classroom. *Teaching English in the Two Year College, 35*(3), 272–281.

Noble, S. U. (2018). *Algorithms of oppression: How search engines reinforce racism*. NYU Press.

O'Neil, C. (2016). *Weapons of math destruction: How big data increases inequality and threatens democracy*. Broadway Books.

Rogers, R. (2011). *An introduction to critical discourse analysis in education*. Routledge.

de Roock, R. S. (2020). On the material consequences of (digital) literacy: Digital writing with, for, and against racial capitalism. *Theory Into Practice, 60*(2). doi: 10.1080/00405841.2020.1857128

de Roock, R. S., & Baildon, M. (2019). MySkillsFuture for students, STEM learning, and the design of neoliberal citizenship in Singapore. *Cognition and Instruction*, 37(3), 285–305.

Scollon, R. (2002). *Mediated discourse: The nexus of practice*. Routledge.

Scollon, R. & Scollon, S. W. (2004). *Nexus analysis: Discourse and the emerging internet*. Routledge.

Scollon, S., & de Saint-Georges, I. (2012). Mediated discourse analysis. In J.P. Gee & M. Handford, *The routledge handbook of discourse analysis*, 66–78. Oxfordshire: Taylor & Francis.

Song, L., Singleton, E. S., Hill, J. R., & Koh, M. H. (2004). Improving online learning: Student perceptions of useful and challenging characteristics. *The Internet and Higher Education*, 7(1), 59–70.

Tyack, D. B., & Cuban, L. (1995). *Tinkering toward utopia*. Harvard University Press.

Van Dijk, T. A. (1998). *Ideology: A multidisciplinary approach*. SAGE.

Weiss, R. E. (2000). Humanizing the online classroom. *New Directions for Teaching and Learning, 2000*(84), 47–51.

7

THE TIPPING POINT IN THE PLATFORMISATION OF DUTCH PUBLIC EDUCATION?

How to approach platformisation from a values-based perspective

Niels Kerssens and Mariëtte de Haan

Introduction

In the Netherlands, digital platforms – computational infrastructures that shape user interactions (Poell et al. 2019) – have become increasingly prevalent in public education in the past decade (Kerssens and Van Dijck 2021). Personalised and data-driven learning environments are being developed as networked constellations which interconnect learning tracking systems, adaptive learning platforms, access portals, learning analytics, dashboards, edu-app packages and cloud-based services of Big Tech companies (ibid.). This platformisation of education in the Netherlands, as in other places in the world, unfolds as a transformative process reshaping education (Decuypere et al., 2021) through pedagogies encoded into digital platforms (Perotta et al., 2020; Williamson, 2017) and through platform architectures of datafication by which social practices, such as teaching and learning, are regulated and governed through platforms' systematic collection, algorithmic processing, and circulation of data (Van Dijck et al., 2018; Van Dijck and Poell, 2017).

Yet as we will show, interconnected processes of platformisation and datafication, at least in the Netherlands, not only *reshape* education but are also *shaped by* already existing cultural practices and rationales (Poell et al., 2019). Considering platformisation from this angle necessitates a thorough examination of what rationales underpin the integration of digital platforms in education, how these uniquely shape this process, while also opening up possibilities for how platformisation could or should be shaped otherwise. Therefore, we will start by thinking historically about platformisation and datafication in Dutch public education, following media scholar David Beer's historical approach to big data (Beer, 2016) to develop a richer understanding of historical discourses and rationalities currently shaping the integration of digital platforms into public schools.

DOI: 10.4324/9781003136842-7

In the following section we uncover how the integration of digital platforms in public education in the Netherlands is part of at least two decades of educational reforms grafted onto *instrumental rationality* – with technologies and techniques of datafication viewed as core instruments for maximising learning effectivity and performance. In the Netherlands, as we will show, digital platforms – presenting more advanced tools of datafication and personalisation – have the potential to seamlessly, unquestioned, and almost naturally 'land into' existing justifications and reasoning around the adoption of technologies in education.

Thereafter, we map how a counter-discourse emerges from public stakeholders and interventions in the educational field, which rallies against an instrumental rationale for digitalisation of educational practice, whilst communicating a clear vision on how digital platform technologies should be taken up through a values-based perspective. This budding counter-discourse, we argue, presents a tipping point which can be further sustained through interaction with earlier fundamental critiques by educational scientists that have been voiced against an instrumentalist and technocratic uptake of educational technology. Finally, in the conclusion we reflect on what our analyses might mean in terms of advice and guidelines for policy makers and practitioners.

The performance-based approach, instrumental logic and the push of platformisation in Dutch public education

What logic underpins and pushes platformisation of school learning environments in the Netherlands? To answer this question, we deconstruct a model of professional action for educators known in Dutch as 'opbrengstgericht werken' ('the performance-based approach'), perceived and actively promoted by the Dutch government as a means to improve learning performance (MECS 2007, 2011). To expose its ideological underpinnings, we will trace the roots of performance-based working to the instrumental rationale of New Public Management or NPM (Gunter et al., 2016) and the 'evidence-based movement' (Eryaman and Schneider, 2017), which have affected educational reforms internationally. From NPM the performance-based approach inherited a dataist style of instrumental and managerial thinking, in particular its view of datafication – the systematic registration, tracking and analysis of data about learners and learning – as a key *instrument* for gaining insight in, controlling and maximising learning processes and learning performance. The evidence-based approach, in turn, nurtured a restricted instrumental view of digital technologies' educational value, investigating educational means – including digital technologies – exclusively in terms of their effectiveness (Clegg 2005; Clemens, 2018). Providing new technological possibilities for datafication and personalisation, new digital platform technologies, we argue, landed comfortably in a performance-based model of professional action, in search of tools for governing and improving learning processes.

Performance-based working was first introduced by the Ministry of Education, Culture and Science (2007) in response to the assumed declining performance of Dutch students in the core subjects as evidenced in international comparisons such as PISA, PIRLS and TIMSS. It replaced the earlier 'meetgestuurd onderwijs' (internationally referred to as 'data-driven teaching'), which have roughly the same connotation (Ledoux et al., 2009): teachers need to make better use of 'objective' student data, such as test results, to inform their teaching (MECS, 2007). In the national policy agenda 'Schools for tomorrow' (MECS, 2007) performance-based working was presented as a key target point for improving language and math performance of all students in primary education.[1] In general, according to the Dutch government, performance-based working refers to schools which "work systematically and purposefully to maximize the performance of its students" (IoE, 2010, p.4; Doolaard, 2013 in Faber et al., 2015). This also means that schools systematically collect data about pupils' learning progress through continuous assessment, interpreting assessment data to enhance learning processes of individual students (MECS, 2007).

In performance-based doctrine, 'objective' student data on student performance is viewed as instrument to inform decision-making by education professionals with a final goal of achieving higher learning outcomes. This characterises its instrumental rationality, which in general refers to a logic of means-to-end thinking in which means, or instruments, are recognised and adopted in terms of their efficiency and effectiveness in achieving particular ends (Kolodny and Brunero, 2013). Such instrumentalist perception of data roots in the output orientation of NPM's managerialist philosophy, which promoted the idea that judgements about performance are most effective when based on outcome data (Gunter et al., 2016). With NPM reforms of education, then, extensive and continuous data collection, realised through different instruments and technologies of measurement (Williamson et al., 2020), becomes central to educational governance (Ozga, 2009) and the improvement of learning processes (Thoutenhoofd, 2018). This is clearly shown in the Dutch performance-based approach. In the doctrine of performance-based working, we recognise this managerial shift towards output and delivery and the central role assigned to datafication as key instrument for optimising learning and educational management; converting learning results into data, make them susceptible to human interpretation and computer processing, to make judgements about learner performance, while also making management decisions based on learners' performance data and related school quality indicators.

In Dutch education, the New Public Management rationale was a fertile ground for another marriage to make education more *effective*: that of education and the educational sciences. While NPM originates from theories in bureaucracy and administrative reform doctrines (Hood, 1991), the evidence-based rationale comes from a marriage between science and policy in which politicians base their policies on 'what works' (Biesta, 2007). And, while New Public Management thinking was a fertile ground for establishing datafication as key instrument for performance-based working, the evidence-based doctrine pushed its acceptance and roll-out by delivering the 'means' – the didactic designs and interventions, the missing link in an outcome based and means-end thinking – for which proper evidence about effectiveness existed from scientific research. Performance-based policy and rationale was 'evidence-based' from the start, based on the conviction that research evidence should be its foundation, even if this conviction has been debated since it was launched (Gravemeijer & Kirschner, 2007).[2] Nonetheless, performance-based working and its operational logic of datafication, were pushed as instruments for enhancement of learning based on scientific claims evidencing it enhanced student performance (EC, 2012).

In the Netherlands, especially in primary education, the datafication of learning took the form of a so-called 'learning tracking system' (LTS), which the Dutch educational council (EC, 2011) and Ministry of Education portrayed as "indispensable *tools* [emphasis added] for schools to work in a performance-based way" (MECS, 2011). The national education council viewed LTS as instruments of (1) *datafication*, "a coherent system of standardized [assessment] instruments and procedures to systematically monitor student performance and development" (EC, 2011) and (2) *personalisation*, "[tools for] tailoring education to the individual level of competence of pupils" (EC, 2012). LTS contained standardised tests for registering and collecting assessment data of individual pupils for at least the subject's Dutch language and Maths. In most cases, however, it also collected data about students' socio-emotional development and learning and behavioural problems (EC, 2011). To push performance-based working in schools, the Dutch government legally obliged the use of LTS in primary schools by amendment to the Dutch primary education act in 2013, making it mandatory to collect individual data about learning progress and results, health and support need (MECS 2021), although schools were free to decide which student tracking system to use (EC, 2012).

In the Netherlands, learning tracking systems were introduced in primary education already in the late 1980s as instruments for systematically tracking and recording student progress (Gillijns and Verhoeven, 1991). Initially, these systems were fully paper based, yet in the 1990s they were increasingly supplied with a computer program in which learning data could be manually and automatically registered, such as the LTS provided by CITO, the Dutch organisation for developing and administering exams and tests. After the turn of the millennium, paper systems were progressively replaced by online digital tracking systems, developed by commercial software companies. These digital web-based systems, such as ParnasSys (Topicus) and Esis (Rovict), incorporated instruments of datafication and personalisation to gradually transform into digital platforms; full-fledged computational data infrastructures which revolved around the systematic registration, algorithmic processing and circulation of data about learners and learning. Their use grew exponentially in the past decade, with ParnasSys as the undisputed market leader; in 2012 60% of primary schools already worked with ParnasSys (Vogelaar, 2012) and in 2016 it had a market share of 60–80% (Bisschop et al., 2016).

The acceptance and legitimation of these platforms of datafication as *tools* for performance-based working commenced further platformisation now with more advanced and more integrated technologies. LTSs are increasingly integrated with digital products and services – adaptive learning platforms, dashboards, learning analytics – presenting new technological possibilities for realising data-driven and personalised learning at a low cost and with a greater impact (Kerssens and Van Dijck, 2021). Importantly, these more advanced digital platform technologies landed comfortably within an already established performance-based model, in search for more effective tools of datafication and personalisation to improve learning processes. This development enabled new research agendas in educational science within the evidence-based approach. The interest in investigating adaptive learning platforms in terms of their *effectiveness* as instruments for improving learning achievements and motivation (Faber and Visscher, 2016; Kester et al., 2018; Molenaar et al., 2016) seems to be based on a performance-based view of what value digital technologies present for education.

Finally, it should be said that the adoption of performance-based working was partial, and its implementation to a certain extent was not successful (Ledoux et al., 2009; IoE 2010) as schools and teachers did not fully leverage the potential of data and other technical instrumentation (Faber et al., 2015; Heemskerk et al., 2014). Despite this, its instrumental rationale clearly was considered 'mainstream' and can therefore still be seen as a dominant force shaping further platformisation of education, given the fact that its desire for effective instruments aligns so neatly with the promises of datafication and personalisation associated with educational platform technology.

The emergence of a counter-discourse for educational platformisation

In past years, a counter-discourse has emerged in which public stakeholders and interventions from public organisations draw on critical media studies and critical pedagogy perspectives to rally against educational platformisation unfolding through an instrumental rationale, while also putting forward a grammar of alternative rationales for digitalisation of education. Even if much is still lacking in the practical realisation of such enterprise, these interventions stand out for their clear vision on how digitalisation of public education should be taken up and for its provision of alternative principles of how this should and can be done. These critical voices are first and foremost based on increasing awareness about platformisation redirecting organisational power from schools to (big) tech companies and acknowledging the role an instrumental model for implementation of these technologies plays in this process.

Dutch public sector organisations Kennisnet and the PO-council, supported by the ministries, are on a mission to reclaim *public control* over educational platformisation, and thereby reclaim control over educational digital reform by private companies. This mission is powered by a new national policy agenda for digitising public education, which lists "sustained focus on the ethics of digitisation in education" as one of five key points, ambitions and activities which should steer digitisation in primary and secondary education for 2019–2023 (MECS 2019). The main trigger for this growing critical awareness of the impact digitisation may have on education are media studies critiques of platformisation as a process of educational reform through privatisation which risks challenging the interests and values of education as a public good (Van Dijck et al., 2018). These scholarly critiques have received much attention in the educational field in the Netherlands, which is far from surprising as in the past years digital platforms, services and infrastructures of (big) tech corporations have increasingly penetrated school learning environments (Kerssens and Van Dijck, 2021).

Within this context of educational platformisation, critical interventions react against two of the instrumental models' governing principles: (1) its vision on educational technologies as objective tools is replaced by a view of technology as a non-neutral and organisational power steering education; and (2) its belief in platforms as tools serving instrumental values of performance and effectiveness is replaced by an encompassing values-based notion of platformisation as affecting public, social and pedagogical values of education. In the rest of this section, we describe how a counter-discourse is shaped through two critical interventions grouped together in the toolkit 'Realising values in digitisation': *The Ethical Compass* (Kennisnet, 2019) and *The Bloom workshop* (Studio Monnik, 2020).[3] In addition, we discuss associations between this emerging counter-discourse around educational platformisation and earlier critiques by educational scholars of the instrumental model, voiced in relation to both critiques of New Public Management thinking in education, as well as in relation to the so-called evidence-based paradigm.

The Ethical Compass and The Bloom workshop

The Ethical Compass (2019) is an online conversation tool in the form of a roadmap for digitalisation for primary, secondary and vocational education developed by Kennisnet, an institute financed by the Ministry of Education, but governed by representatives of the educational sector. The tool is aimed at schools and educational professionals to assist them in systematically exploring and answering ethical issues surrounding digitisation through a values-based perspective in small groups of 3–7 professionals, supported by a moderator. Ethical reflection starts by naming and describing the schools' core values, often a mix of public, personal, ideological and pedagogical ones (e.g. autonomy, collaboration, trust, honesty, equality). Subsequently the group formulates an ethical question, such as: Is it good that our school uses technology for personalised learning? Is it good that our school knows everything about our students to help them getting a degree? After first collecting intuitive responses, the group will formulate pro and counter arguments by reflecting on what values are enforced and threatened in the use of a particular educational technology. Collected arguments are weighted and the group finishes by formulating the answer.

The Bloom (online) workshop (2020) is a conversation tool for primary and secondary education, developed by Studio Monnik – a company that designs future scenarios to support strategic decision-making by both public and private organisations – and commissioned by the PO-council and the VO-council: the sectoral organisations that represent school boards of all types of public primary and secondary education. The workshop is aimed at school boards to support them to develop a shared vision of digital education aligned with their desired educational values and beliefs. To ignite conversation and vision formation, the workshop is designed as 'social science fiction' (Selwyn et al., 2020) around a provocative future scenario of platformised education in which education has completely moved to the digital environment of the fictional Bloom platform.

Assisted by a virtual human guide, and supported by videos, the participants "travel" to a near future of education in 2035 to explore what impact current technological developments – rise of personalised learning, integrated digital learning environments, learning analytics, AI emotion reading – may have on education. This future of platform-based learning is visualised and narrated in video animations, through which participants follow primary school student Zoey's everyday interactions with Bloom. In their journey, participants are triggered to reflect on their vision of technology in education and engage in conversations on possible questions about the impact platforms may have on education: If education will soon take place via a platform, will there still be meaningful contact between teacher and student? How will artificial intelligence affect the way we judge student development? What if you don't want or can't afford education in Bloom, are there any alternatives? At the end of the workshop, the participants formulate a joint statement as the basis for developing a vision on educational digitisation.

Interventions such as the Ethical Compass and the Bloom workshop aim to strengthen participation by educational professionals in decision-making around the uptake of digital technologies by opening a space for critical reflection and conversation on the shaping powers of educational technologies, and by having educational professionals reflect on their potential impact on educational practices and associated values. In the report "Weighing Values" (Pijpers et al., 2020) – a vision document that introduces and motivates the perspective of the ethical compass – Kennisnet writes, "if you regard digitisation as neutral, as a development that is inevitable [...] then you forget that you are also able to steer it". Schools and educational professionals are portrayed as agents who can actively reflect on what values are enforced as well as endangered. As such, these interventions intend to show that schools are able to exert control over the application of digital technology and "indeed have something to say about the use of technology" (ibid.). Rather than focusing on questions such as "is this technology effective?" or "does this technology contribute to improve student performance?", they encourage educators to ask questions about ways in which digital technology works with and/or against their own educational values and principles. For example, "what happens to meaningful teacher-student contact when digital learning systems take over parts of the teacher's work?" and "what does that mean for the teacher's professional autonomy?". Bringing conversations around these questions into schools, these interventions aim to offer schools a starting point for reclaiming control over shaping digital education.

Moreover, both interventions advocate a values-based perspective, encouraging educators to steer digitalisation by positioning public, social and pedagogical values of education upfront rather than only instrumental values, such as effectiveness. The

Ethical Compass calls for schools and decision makers to not only focus on what is technologically possible, but to consider the social consequences of educational technologies in terms of their impact on public educational values. In practice this means the tool encourages educators to reflect on how particular technologies may affect the professional autonomy of teachers, equality of opportunity and meaningful interactions between teacher and pupils. This 'ethical perspective' – thinking about doing the right thing and reflecting on public values – challenges instrumentalist approaches which still play such a dominant role in shaping what and how digital platforms integrate in public education.

Earlier critiques against an instrumentalist uptake of technology

This budding counter-discourse resonates with earlier fundamental critiques by educational scientists and educational practitioners raised against an instrumentalist and technocratic uptake of educational means, strategies and techniques which ignore the values-orientation of education, that is, the idea that education always revolves around making decisions that imply certain values. Particularly relevant here are earlier critiques by educational scholars in response to New Public Management thinking as well as evidence-based philosophies. While earlier in this chapter we have shown that New Public Management and evidence-based philosophies are the historical roots of a current logic of platformising education, in this section we present earlier critiques on both these philosophies, as we believe this helps to position the recent counter-discourse into a larger discussion on the adoption of technology in education. Such synthesis also provides additional practical insights into the task of digitising public education. These earlier critiques relate both to the point that technology has organisational power that affects educational values, and to the reaction in the described counter discourses that we need values-based rationales for the adoption of technologies in education. However, they add important dimensions to this argument which takes into account the nature of education and learning, as well as the issue of autonomy of educational professionals in processes of platformisation and datafication.

Earlier debates suggested that an instrumental rationale in the adoption of technology not only implies the *absence* of a values-based rationale, but also imposes a *model for professional action* in which the professional is not in control of the desirable outcomes while being 'on the job'. Particularly relevant here is the argument of Gert Biesta (2007, 2010a, 2010b), although this point has been made across the field of education (for instance, Wubbels and van Tartwijk, 2017; Gravemeijer & Kirschner, 2007). Biesta has raised concerns in the context of his critique of evidence-based research in which "it is assumed that the only relevant research questions are questions about the effectiveness of educational means and techniques" (2007, p. 5). As Biesta describes, such an evidence-based approach implies an instrumental model for professional action, in which 'the end' is already given and the only question for the professional is how to reach this end as efficiently as possible, while not being able to judge the desirability of educational ends. Biesta warns us against a mechanistic view of teaching in which it is the job of 'others' (be it educational scientists, or programmers of educational software) to program and automatise teacher activities based on earlier 'evidence' of effectivity.

Importantly, Biesta (2007) has argued that an instrumental model is not just 'not suitable' to encourage educational professionals to take up their roles to consider the social and moral consequences of educational technologies, but that it also actively discourages such roles. An instrumental rationale, according to Biesta, not only 'restricts the scope of decision-making to questions about effectivity and effectiveness' but also 'restricts the opportunities for participation in educational decision-making' (p.6). Within an instrumental rationale focused on the effectiveness of educational technology, it is assumed that decision-making about adopting these 'instruments' in educational practice is a procedural issue, thus morally 'neutral'. Such concerns around the reduction and obscuring of decision-making processes are all the more relevant in platform technologies organised through algorithmic processing which 'blackbox' (pedagogical) decisions in platform design for educational professionals. A legitimising rationale for the adoption of educational technologies based on an effectivity discourse further obscures the ethical choices implied in such processes.

This argument that an instrumental rationale has organisational power and implies a model for professional action has also been addressed in critiques on NPM thinking in education. The specific point that we like to bring in from this earlier NPM related discussion in addition to the new counter discourses against an instrumental approach, is that data technologies have the tendency to shift normative decision-making away from professionals 'on the work floor' to other levels of the organisation, which then further limits the autonomy of these professionals. In other words, such processes have 'system consequences' for accountability and responsibility and on the location of *where* in the system pedagogical decision-making takes place. This point adds a 'system and labour division point of view' to the new counter discourses, which we think is key to consider in designing strategies for platformisation in education. Gunter et al. (2016), for instance, have warned that according to an instrumentalist, NPM based thinking, intervention power and accountability is moved from pedagogy 'itself' towards output registering technologies and management staff. Such labour division shifts have been a concern in the educational sector for some time, while warning for a shift from professional norms and procedures to management targets and outcomes (Edwards and Daniels 2012), or for the danger that democratically legitimised bodies of supervision can come under threat when control is moved away from classrooms and schools to other systems of accountability (Hangartner and Svaton, 2013). Such a system view with an eye for labour division effects and its concurring shifts in accountability and responsibility are particularly relevant when bearing in mind that platform technologies introduce even more radical shifts in the organisation of labour. For instance, Perrotta et al. (2020) describe how through the use of Google Classroom, key responsibilities, such as institution level integrations, device management, the choice of data region policy and home access, are transferred to system administrators.

As a possible answer to a technologically determinist interpretation of changing models of professional action, Biesta advocates an agentic model in which professionals should be encouraged to hold on to what he calls an "ends-in-view" attitude (2007), irrespective of what technology is introduced in their educational practice. Biesta argues that decision-making in teaching, includes always also 'on the spot' and 'in context' judgements of 'what is desirable or appropriate', irrespective of whatever knowledge, tools, methods are launched to assist. As Biesta writes, interventions of teachers are always in some way interventions in the 'existing course of events' which are then opportunities for students to respond to and serve a mutual learning process of both teachers and students. In Biesta's agentic model of professional action, technologies can inform professional action, when not used as a recipe, and when keeping an eye on the uniqueness and context of each teaching learning action. Relevant for platformisation, just as for the adoption of any technology in education, is that professionals should not be enslaved by pre-established ends already implied in technologies but steer their actions towards what they consider desirable ends, informed by what technologies can offer them. We believe that his agentic model of professional action, as well the argumentation laid out above about why it is essential, further sustains the counter discourses on platformisation and the possibilities to steer platformisation in a practical sense.

Conclusions: what is needed to sustain the mission of a values-based platformisation?

Informed by our analyses of an instrumental rationale underpinning platformisation of public education in the Netherlands, its history in earlier practices of datafication and associated new public management and evidence-based philosophies, our analyses of recent counter discourses and connections to earlier critiques by educational scholars on instrumental reason, we will now move towards the more practical question: *what is needed to sustain the mission of a values-based platformisation?* What do policy makers, sector representatives, publishers, ICT and software designers, schools and teachers need to know? And what do they need to do, given the challenges of platformisation we have brought up?

We believe the Dutch case might be exemplary for how digital education can be shaped under public control through a values-based perspective in the pre-implementation phase of digitisation (rather than a later phase of already platformised schooling). This pre-implementation phase is key for schools and teachers as the shape of the digital learning environments they will work in is affected by many of the decision-making taking place within this phase. Moreover, once particular technologies are adopted such decisions are hardwired into the set-up of digital learning environments and cannot easily be changed.

With the *Ethical Compass* (2019) and *The Bloom workshop*, the toolkit 'Realising values in digitisation' offers hands-on tools which schools and educational professionals can use to make a start with implementing digital technologies through a values-based perspective that privileges public, social and pedagogical values of education over instrumental ones. By stimulating school leaders and teachers to reflect on how digital technologies may impact desired educational practices and associated values, we believe interventions such as these can play an important role in strengthening their control over how digital platforms shape public education.

Yet to sustain this mission, we argue, more is needed. In the pre-implementation phase, which should also include the pre-design phase, the conditions for digital and platform-based education should be discussed and debated not only by school leaders and teachers, but by a variety of different stakeholders shaping digital education, including policy makers, representatives of the various sectors of public education, educational publishers and software developers. Based on key insights from our analyses, we will now present three points of attention which all aforementioned stakeholders that play and should play a role in decision-making in the pre-implementation phase needed to account for to steer educational platformisation through a values-based orientation prior to the design and assumed massive implementation of platform technologies by schools:

1. ***Historical awareness and analyses of existing rationales for the adoption of technology is key to design 'new' rationales for platformisation.***

Critical research into educational platformisation points out that digital platforms, their infrastructures and encoded pedagogies, reshape education and challenge education as a public good. Yet, equally important, we argue, is to find out what already established cultural rationalities and practices of education facilitate the unquestioned integration of digital platforms in education, how its power dynamics uniquely shape platformisation, and how it could or should be shaped otherwise. As we discussed in this chapter, the platformisation of primary school education in the Netherlands was built on and pushed by an instrumental rationale for adopting technology in education. To think historically about platformisation of education as a process that is part of an earlier history of educational reforms and associated ideologies and philosophies, is also important for professionals and policy makers directly involved with decision-making about digital platform technologies in schools.

However, historical awareness about the process of implementation implies that in addition to 'just' thinking about what the 'right' values or the right rationale for platformisation are in a present context, educational professionals need to actively address (e.g., contradict, deal with, come to terms with, understand its origin) those of the past which are still affecting the present. Historical awareness and an analysis of existing rationales for the implementation of educational technologies, and analyses of impact on existing educational practice, levels of accountability and autonomy, and its implicit pedagogies, should guide platformisation at all stages of implementation, next to putting desired values upfront. In practical terms, this might mean that educational professionals and policy makers not only systematically explore and address ethical issues surrounding digitalisation and ask what core values they think should steer this process, as is recommended in the Ethical Compass, but also make an analyses of their existing ideologies of the adoption of technology, how these impact upon their core educational values, and what alternative positioning towards digital technologies follow from such analyses.

2. ***For values-based platformisation of education it is key to also design a corresponding model of professional action in which professionals are able to act in an agentic way with technologies.***

In order to encourage platformisation to be based on public, social and pedagogical values of education, it should always allow that an 'ends-in-view' perspective is enabled by educational professionals, and the legitimate authorities when decisions about the design of educational means and 'what desirable education is' is at stake. This does not only imply that educational professionals and representatives should take part in the decision-making process related to which educational values should guide the design of technology. It also implies designing platform technology with enough 'degrees of freedom' for professionals to intervene in them, and that values related decisions implied in already designed platform technology and software are made explicit, especially given that much of the technology is 'blackboxed' for professionals.

In practice, this means that policy-makers, sector representatives, publishers, ICT and software designers need to ask themselves what the appropriate 'entry points' are where professionals can access and act with data or actively manipulate part of the technology. In other words, they should ask how much 'white boxing' (Säljö, 2012) do professionals and students need in order to work with the platform so that they have enough agency? What kind of training (platform or data literacy) is needed and effective in this respect? The answers to these and similar questions will form what we have called a *model for professional action*, and we think policy makers, sector representatives, publishers, ICT and software designers need to be informed that for platform technology to function properly in the field of education, they should not just design technologies, but design a model of professional action, in close alignment with the design process.

3. ***A system perspective point of view is needed when keeping track of and steering values-based platformisation.***

Lastly, based on our earlier point that data technologies have the tendency to shift normative decision-making away from professionals 'on the work floor' to other levels of the organisation, we argue that a 'system and labour division point of view' is needed when platform technologies, their accompanying values framework and model of professional action, are designed and implemented. Platform technologies induce new divisions of labour in the educational process between professionals and automatised technologies, but also between different professional categories (such as between teachers, school management and those responsible for software development, and for data and technology management). Here again, the issue is how in such new divisions of labour professional autonomy and an 'ends-in-view perspective' of educational professionals, as well as 'continuous democratic contestation and deliberation' (Biesta, 2007) can be best shaped.

Policy-makers, sector representatives, publishers, ICT and software designers thus will have to operate at, design for, include and create collaborations that address the entire system involved in the creation, dissemination, implementation and use of these technologies. This new labour division asks for a different perspective on decision-making in the implementation of digital technology in education and a critical look at where responsibility, autonomy and accountability lays. All stakeholders will have to realise that this necessitates new collaborations and meeting points between software developers, publishers, educational boards, sector representatives and policy-makers in which these issues are scrutinised and in which educational professionals will claim authority over decisions related to pedagogical values.

One way of doing this is the creation of sector wide cooperatives that are assigned authority for decision-making or the advice of authorities. Examples in the Netherlands which have already been launched are Edu-K and SIVON. Edu-K is a public-private cooperative of educational publishers, suppliers, software developers and umbrella organisations of schools, which has the lead in designing the preconditions for the use of ICT in education to the benefit of public schools (Edu-K, 2021). SIVON is a cooperative for joint tendering for ICT-products and services for public education (SIVON, 2021). These cooperatives might serve as a model for what new labour divisions might look like. Even if their agendas are dominated by themes of privacy and security and issues collective purchasing, they may just as well provide examples of how private and public stakeholders can further discuss and take responsibility for, and authority over, shaping the values-based pedagogies of platformisation.

To conclude, in the past ten years, as we demonstrated in the first part of this chapter, the platformisation of primary education in the Netherlands developed as part of an already existing rationale built on instrumental values of efficiency, effectiveness and performance. Digital platforms and services have been integrated into learning environments motivated by their expected contribution to the datafication and personalisation of learning, and overall to the innovation and improvement of education. Digitisation, obviously, offers advantages, and it is key for educational scholars to investigate how these technologies can make important contributions to improving education and learning performance. At the same time, platformisation challenges education as a public good which issue forms a key incentive for schools, supported by public sector organisations and the ministries, to reclaim *public control* over educational platformisation through a values-based orientation. The three points of attention described above aim to give further direction to this mission.

Notes

1 Although we lack data to provide a full picture how wide and how thorough this policy was adopted, there is evidence that not all of the schools have adopted the performance-based approach as it was designed. For instance, in 2011, 30% of Dutch primary schools

scored satisfactorily on the inspection indicators for performance-based working; even though the aim was to double this in 2015 and increase it to 90% in 2018 (MECS 2011).

2 In the Netherlands, the evidence-based approach was promoted by the Committee Parlementary Educational Research (Commissie Parlementair Onderzoek Onderwijsvernieuwing) which concluded that educational innovations in the Netherlands were not (sufficiently) based on educational research and pleaded to base future educational innovations on empirical evidence.

3 Also, part of the toolkit is the 'Value Framework', which is currently only available in beta-version and will be finalised in 2021. The framework identifies public values underpinning Dutch education and is configured around three core values – justice, humanity, autonomy – that branch further into underlying values such as equality, inclusivity, meaningful contact, student self-determination, freedom of education, etc. Based on collectivity and commonality, the value-framework is intended to enhance public control over digitisation, providing the sector of public education with a common language by which to steer market developments.

Bibliography

Beer, D. (2016). How should we do the history of Big Data? *Big Data & Society*, *3*(1). doi:10.1177/2053951716646135

Biesta, G. 2010b. Why 'what works' still won't work: from evidence-based education to value-based education. *Studies in Philosophy and Education*, 29(5), 491–503. doi: 10.1007/s11217-010 9191-x.

Biesta, G. J. J. (2007). Why 'what works' won't work: Evidence-based practice and the democratic deficit of educational research. *Educational Theory*, 57(1), 1–22.

Biesta, G. J. J. (2010a). *Good education in an age of measurement: Ethics, politics, democracy*. Boulder, Co: Paradigm Publishers.

Bisschop, P., Imandt, M., Van der Vegt, A. L., & Bomhof, M. (2016). *Kwaliteit, betrouwbaarheid of innovatie: onderzoek naar de leermiddelenmarkt in het primair onderwijs.*

Clegg, S. (2005). Evidence-based practice in educational research: A critical realist critique of systematic review. *British Journal of Sociology of Education*, 26(3), 415–428, doi: 10.1080/01425690500128932

Clemens, W. (2018). *Evidence and its integration into teacher knowledge: Foucaultian perspectives to link research knowledge and teaching. Journal of Education for Teaching. 44*(5), 637–650. doi: 10.1080/02607476.2018.1516352

Decuypere, M., Grimaldi, E., & Landri, P. (2021). Introduction: Critical studies of digital education platforms. *Critical Studies in Education*, *62*(1), 1–16. doi:10.1080/17508487.2020.1866050

EC (Education council). (2011). *Toetsing in het primair onderwijs.*

EC (Education council) (2012). *Advies Wetsvoorstel LVS.*

Edu-K. 2021. "Over Edu-K." *Edu-K.* Accessed 15 April 2021. https://www.edu-k.nl/.

Edwards, A. & Daniels, H. (2012) The knowledge that matters in professional practices, *Journal of Education and Work*, *25*(1), 39–58, doi: 10.1080/13639080.2012.644904.

Eryaman, M. Y., & Schneider, B. (Eds.). (2017). *Evidence and public good in educational policy, research and practice.* Springer International Publishing

Faber, J., & Visscher, A. (2016). *De effecten van Snappet: Effecten van een adaptief onderwijsplatform op leerresultaten en motivatie van leerlingen.* Enschede: Universiteit Twente.

Faber, J., Visscher, A., & Schut, W. (2015). *Opbrengstgericht werken in het primair onderwijs: competenties, uitvoering en resultaten.*

Gillijns, P., & Verhoeven, L. (1991). Naar een leerlingvolgsysteem voor het basisonderwijs. *Pedagogische Studiën, 68*, 1–15.

Gravemeijer, K.P.E. & Kirschner, P.A. (2007) Naar meer evidence-based onderwijs? *(Towards more evidence-based education?) Pedagogische Studiën*, 84(6), 463–472.

Gunter, H. M., Grimaldi, E., Hall, D., & Serpieri, R. (2016). *New public management and the reform of education: European lessons for policy and practice*. Routledge.

Hangartner, J. & Svaton, C.J. (2013) From autonomy to quality management: NPM impacts on school governance in Switzerland. *Journal of Educational Administration and History*. 45(4), 354–369, doi: 10.1080/00220620.2013.822352

Heemskerk, I., Verbeek, F., Kuiper, E., Oomens, M., van der Linden, J., & Hilbink, E. (2014). 28 Opbrengstgericht-werken-in-het-voortgezet-onderwijs. Beeld en beleid. RAPPORT: 915 ISBN: 978-90-6813-976-1, https://kohnstamminstituut.nl/rapport/opbrengstgericht-werken-in-het-voortgezet-onderwijs/.

Hood, C. 1991. A public management for all seasons. *Public Administration*, 69(Spring), 3–19.

IoE (Inspectorate of Education). (2010). *Opbrengstgericht werken in het basisonderwijs: Een onderzoek naar opbrengstgericht werken bij rekenen-wiskunde in het basisonderwijs*. Utrecht: Inspectie van het onderwijs

Kennisnet. 2019. *The Ethical Compass*. Accessed October 12 2021. https://wijzer.kennisnet.nl/ethiekkompas

Kerssens, N., & Van Dijck, J. v. (2021). The platformization of primary education in The Netherlands. *Learning, Media and Technology*, 1–14. doi:10.1080/17439884.2021.1876725

Kester, L., Amina Cviko, Caressa Janssen, Mario de Jonge, Monika Louws, Suzan Nouwens, Tineke Paas, Frauke van der Ven, Wilfried Admiraal, Lysanne Post, Ditte Lockhorst, Michael Buynsters, Geertje Damstra. (2018). *Docent en leerling aan het stuur: Onderzoek naar leren op maat met ict*. Retrieved from https://www.nro.nl/wpcontent/uploads/2018/02/Doorbraakproject-Onderwijs-ICT_Eindrapport.pdf

Kolodny, N., & Brunero, J. (2013). Instrumental rationality. In E. N. Zalta (Ed.), *The Stanford encyclopedia of philosophy* (Fall 2013 ed.). Retrieved from http://plato.stanford.edu/archives/fall2013/entries/rationality-instrumental/

Ledoux, G., Blok, H., Boogaard, M. Met medewerking van Krüger, M. (2009). Opbrengstgericht werken; over de waarde van meetgestuurd onderwijs. SCO-Kohnstamm Instituut van de Faculteit der Maatschappij – en Gedragswetenschappen, Universiteit van Amsterdam (SCO-rapport nr. 812, projectnummer 40296).

MECS (Ministry of Education, Culture and Science). (2007). *Scholen voor morgen: Samen op weg naar duurzame kwaliteit in het primair onderwijs*. Den Haag: Ministerie van OCW

MECS (Ministry of Education, Culture and Science) (2011). *Actieplan 'Basis voor presteren'*. Den Haag: Ministerie van OCW

MECS (Ministry of Education, Culture and Science). (2019). *Digitaliseringsagenda primair en voortgezet onderwijs*. Ministerie van Onderwijs, Cultuur en Wetenschap

MECS (Ministry of Education, Culture and Science). (2021). *Welke gegevens bewaart De basisschool over mijn kind?* Retrieved April 07, 2021, from https://www.rijksoverheid.nl/onderwerpen/basisonderwijs/vraag-en-antwoord/welke-gegevens-bewaart-de-basis school-over-mijn-kind

Molenaar, I., Campen, C. A. N.V., & Gorp, K.V. (2016). *Onderzoek naar Snappet; gebruik en effectiviteit*. Retrieved from Nijmegen: http://hdl.handle.net/2066/167523

Studio Monnik. 2020. *The Bloom Workshop*. Accessed October 12 2021. https://bloomworkshop.nl/

Ozga, J. (2009). Governing education through data in England: from regulation to self-evaluation. *Journal of Education Policy, 24*(2), 149–162. doi:10.1080/02680930902733121

Perrotta, C., Gulson, K. N., Williamson, B., & Witzenberger, K. (2020). Automation, APIs and the distributed labour of platform pedagogies in Google Classroom. *Critical Studies in Education*, 1–17. doi:10.1080/17508487.2020.1855597

Pijpers, R., Bomas, E., Dondorp, L., & Ligthart, J. (2020). *Waarden wegen: Een ethisch perspectief op digitalisering in het onderwijs*. Zoetermeer: Kennisnet.

Poell, T., Nieborg, D., & van Dijck, J. (2019). Platformization. *Internet Policy Review*, *8*(4), 1–13.

Säljö, R. (2012). Literacy, digital literacy and epistemic practices: The co-evolution of hybrid minds and external memory systems. *Nordic Journal of Digital Literacy*, 01, 5–19.

Selwyn, N., Pangrazio, L., Nemorin, S., & Perrotta, C. (2020). What might the school of 2030 be like? An exercise in social science fiction. *Learning, Media and Technology*, *45*(1), 90–106. doi:10.1080/17439884.2020.1694944

SIVON. 2021. "Onze missie en visie." Accessed April 15 2021. https://www.sivon.nl/over-ons/onze-missie-en-visie/.

Thoutenhoofd, E. D. (2018). The datafication of learning: Data technologies as reflection issue in the system of education. *Studies in Philosophy and Education*, *37*(5), 433–449. doi:10.1007/s11217-017-9584-1

Van Dijck, J., & Poell, T. (2017). Social media platforms and education. In Jean Burgess, Alice Marwick & Thomas Poell (Eds.), *The SAGE handbook of social media* (pp. 579–591). London: Sage.

Van Dijck, J., Poell, T., & Waal, M. d. (2018). *The platform society: Public values in a connective world*. New York: Oxford University Press.

Vogelaar, L. (18 September 2012 dinsDag). Klas in kaart. *Reformatorisch Dagblad*. https://advance-lexis-com.proxy.library.uu.nl/api/document?collection=news&id=urn:contentItem:56KV-K701-JC8W-Y1JY-00000-00&context=1516831

Williamson, B. (2017). Decoding ClassDojo: psycho-policy, social-emotional learning and persuasive educational technologies. *Learning, Media and Technology*, *42*(4), 440–453. doi:10.1080/17439884.2017.1278020

Williamson, B., Bayne, S., & Shay, S. (2020). The datafication of teaching in Higher Education: Critical issues and perspectives. *Teaching in Higher Education*, *25*(4), 351–365. doi:10.1080/13562517.2020.1748811

Wubbels, T & van Tartwijk, J. (2017). Evidence Informed Innovation of Education in the Netherlands: Learning from Reforms. In Eryaman, M.Y. & Schneider, B. (Eds.), *Evidence and public good in educational policy, research and practice* (pp. 125–142). Springer.

8

"THE BEATLES WITH THE LOWER SCORE, IT BREAKS MY HEART"

Framing a media education response to datafication and algorithmic recommendations in digital media infrastructures

Jérémy Grosman, Jerry Jacques and Anne-Sophie Collard

Introduction

Digital platforms now mediate most of the actions we take when reading news, watching videos, buying goods, finding routes, chatting to people, etc. The growing influence of some, such as Google, Amazon or Facebook, raises both hopes (e.g., freedom, economy,) and concerns (e.g., competition, surveillance). These tensions propel us to invent better ways to discuss the recent and rapid transformations brought by such platforms. The challenge is both to make sense of the main social and technical processes that shape these digital platforms and to design situations in which their consequences become discussible, for anyone concerned. Thus, with the hope of addressing these issues, we sought to bridge recent research from digital media education, about digital literacy and critical education, with in science and technology studies, about engineering practices or digital infrastructures.

We decided to focus on the recommender systems[1] at the core of most of these digital platforms. They promise to provide, both, a unique algorithmic technique for enabling users to orient themselves amidst the wealth of available information and a unique business strategy to keep users engaged with the platforms' contents and advertisements. They need to be approached, both, as technical infrastructures, which foster interactions between specific users and specific items, and as media interfaces, which partly automate the editorial process. The tension between, on the one side, the opaqueness of their functioning due to technical complexity and business confidentiality and, on the other side, the significance of these systems for the information we exchange and the actions we take, need public debate. The educational challenge lies in making sense of how these social and technical dimensions mingle.

To address this problem, we devised an educational activity, titled "In the shoes of an algorithm", that would enable people to understand and critically discuss the

DOI: 10.4324/9781003136842-8

problems they thought were raised by *recommender systems*. The pen-and-paper game (or simulation), designed in close collaboration with two media educators, offered participants the opportunity to think and act like an engineer, designing a recommender system and, like an algorithm, following the engineers' instructions. The shift in perspective would enable participants, we believed, to better reflect on the socio-technical decisions that shape recommender systems and their experience of them – departing, for instance, from more common understandings of privacy issues or technical progress.

This chapter presents the results of qualitative analyses of the observed and recorded sessions of the educational activity, in order to account for its educational effects on participants. We draw three main conclusions. Firstly, the analysis shows that the activity succeeded in helping participants to articulate important technical features shared by most recommender systems. Secondly, it shows how the educational activity succeeded in enabling participants to critically discuss the social values embedded in the technical infrastructures, even sometimes imagining alternative ways. Thirdly, the overall initiative shows the value of experiential learning approaches for bringing participants to reflect, at one and the same time, about their media practice *and* digital infrastructures.

Media literacy and digital infrastructures

The recommender systems, insofar as they provide "means of representation and communication" indispensable in our daily lives, may fruitfully be approached as *media* (Buckingham, 2019, p.14). The concept of media brings to our attention two important dimensions of recommender systems (Anderson and Meyer, 1988). First, it invites us to think about the specific material *and* symbolic systems that are involved in communication processes. In this case, recommender systems may be thought of as systems, made of data and algorithms, which, filtering the information we exchange, contribute to shape our environment. Secondly, it further invites us to consider the actions people engage in when participating in such communication processes. In this case, recommender systems importantly influence the actions we take, from the information we share to the goods we buy. The educational challenge thus becomes one of bringing people to reflect upon how recommender systems influence their media practices.

The media literacy, which media education initiatives usually seek to foster, has traditionally been defined as people's abilities to "access, analyze, and produce information for specific outcomes" (Aufderheide, 1993, p.6). The approach to media literacy education followed here further stresses people' abilities to "share meaning through symbol systems in order to fully participate in society" (Hobbs, 2010, p.31). Hence, the development of technical competences chiefly interests us insofar as they constitute a prerequisite for developing participants' critical competences: capacities to discuss the roles of media in our lives and to imagine alternative ways of relating to them (Jørgensen and Phillips, 2002; Buckingham, 2019). The educational activity therefore aims at providing participants with conceptual tools that, without turning them into technicians, would enable them to articulate social, technical and informational dimensions necessarily involved in media practices (Fastrez, 2010; Ligurgo et al., 2019).

The media education initiative focusing on recommender systems came across two main challenges. Firstly, the development of critical competences requires an educational activity that is open enough to enable participants to share the concerns they experience about recommender systems (see Section 3). Secondly, the development of technical competences requires an educational transaction capable of bringing participants to understand otherwise abstruse technical matters, so as to be able to reflect upon how recommender systems affect media content and practices (Saariketo, 2014). The value of the educational activity lies in its ability to foster a critical discussion enabling participants to relate personal experiences and collective processes along with technical infrastructures and engineering practices (Gray et al., 2018).

The studies on socio-technical systems or infrastructures (Bijker et al., 1987; Star and Ruhleder, 1996) focus not only on the technical features they share, but also on the social and technical processes that determine their design. The literature on recommender systems allowed us to establish a number of features common to most recommender systems (Ricci et al., 2015; Seavers, 2019). Firstly, engineers use *data* for setting recommender systems and for producing actual recommendations. Secondly, engineers design *algorithms* to move from data to rankings so as to foster interactions between users and items. Thirdly, engineers use *metrics* to assess the recommender systems' performances. This challenged us to show how an appropriate level of technical understanding of *data*, *algorithms* and *metrics* may enable genuine critical discussion.

Data about users (e.g., gender, age, location), items (e.g., genre, date, producer) or interactions between users and items (e.g., views, clicks, likes) are central to most recommender systems. The literature in media education (Pangrazio and Sefton-Green, 2020) has stressed the heuristic value of approaching data as a *text*, a situated collection of symbols that could prompt different "activity and inquiry" from readers. Thus, "data literacy" does not stand so much for capacities to read and write data, but rather for capacities to engage in processes where data plays an important part (Selwyn and Pangrazio 2018; Hardy et al., 2019; Seymoens et al., 2020). In a similar vein, the literature in science and technology studies (Gitelman, 2013; Porter, 1996) approaches *data* as traces that can be interpreted (data necessarily is data *from* something) and manipulated (data necessarily is data *for* something). Here, the educational activity would attempt to bring participants to understand how the data can be used for recommending specific items to specific users.

Algorithms provide engineers with means to turn apparently meaningless inscriptions into valuable information. The detailed computational techniques composing recommender systems cannot be easily conveyed to a general audience – they significantly change from one year to another, or from one system to another, and they mobilise a string of complex computational and statistical operations (e.g., k-nearest neighbour, matrix factorisation, neural networks). Hence, the educational activity would try to convey a general notion of algorithm, avoiding technical considerations regarding recommender systems (Hill, 2016; Jaton, 2017). In a nutshell: algorithms may be defined as *stabilised* and *formalised* computing techniques that may be recognised in a variety of inscriptions (e.g., descriptions, codes, traces) and that may be executed by any agent mastering a well-defined set of operations (e.g., reading, writing, adding). Finally, however, we wanted participants also to imagine how engineers would evaluate the performance of their algorithms.

Metrics provide engineers with ways of assessing the behaviour of their algorithms (Slayton, 2015). The performance of recommender systems is usually evaluated offline by comparing, for each user, the list of predicted interactions (i.e., the list of items with which the user is thought most likely to interact) to the list of observed interactions (i.e., the list of items with which the user has actually been observed to interact). The evaluation essentially rests on metrics: mathematical conventions that enable engineers to measure the difference between predicted and observed interactions (Grosman and Reigeluth, 2019). The recent studies in history and philosophy of quantification have shown that metrics constitute a key site where social and technical norms entangle (Porter, 1996; Desrosières, 1992).

This review of media literacy and science and technology studies' literature have helped us to unpack the educational objectives and challenges of an educational activity focusing on recommender systems. The media literacy literature highlighted the need to help learners to articulate the technical, informational and social dimensions of media (Fastrez and De Smedt, 2012; Jacques et al., 2013; Claes and Philippette, 2020). Here, educational objectives essentially lie in equipping participants with an understanding of the socio-technical systems at stake, so that they would become capable of critically discussing and articulating the problems they encounter in their daily media practices. The science and technology literature brought forward practical propositions combining an attention to objects and practices, a focus on data and algorithms, a concern for technical and social values. Here, the educational challenge could be largely addressed by allowing participants to manipulate data, design algorithms and to discuss metrics.

Method

A design-based approach

The critical approach we adopted towards media education meant we treat participants as equals capable of telling what counts for them as important and capable of suggesting ways of addressing such issues (Rancière, 1983). The inquiry was simultaneously an intervention to enable participants to critically engage with digital platforms, and an investigation about participants experience with recommender systems. Thus, the educational activity, rather than benumbing participants' capacities with massive technical explanations, would instead seek to construct a problematic situation in which participants, would think like engineers and act like algorithms, in order to become capable of critically discussing some of the issues they experience in their daily interactions with recommender systems (Simondon, 1958).

We consequently adopted a design-based method to construct our educational activity (Cobb et al., 2003; The Design-Based Research Collective, 2003; Barab and Squire, 2004; Wang and Hannafin, 2005; Anderson and Shattuck, 2012; McKenney and Reeves, 2013). The core idea of DBR is to bring researchers and practitioners to "depart from a problem and then pursue both knowledge and interventions that address it" (McKenney and Reeves, 2013, p.4). Two characteristics mattered. First, the method consists of several iterative steps, comprising several cycles of inventions and revisions (Cobb et al., 2003) that can be described as "research through mistakes" (Anderson and Shattuck, 2012). Secondly, the method explicitly integrates both experts and practitioners in the research team (Cobb et al., 2003).

The educational activity was initially developed in close collaboration with a youth media education organisation based in Namur called Action Médias Jeunes. From the start, it was envisaged as a one-off activity the association would offer to school or youth centres. The informal character of the educational initiative, performed in the classroom by non-academic actors, enabled us to escape from the stable didactic objectives or systematic learning evaluations that are often required to structure school curriculum (Jacobi, 2001). The design process comprised two main stages. The first stage (Jacques et al., 2020), sought to develop an engaging discussion and to analyse the problems voiced by participants. The second stage, which we discuss in this chapter, sought to assess the educational outcomes of the final activity.

The educational activity

"In the Shoes of an Algorithm" is a two-hour long pen-and-paper activity, played by around 20 people aged 14 plus. The objectives are to develop a recommendation algorithm, to rank videos and to evaluate the different recommendation algorithms. During the first phase, participants are divided into three groups. They are expected to act as engineers designing a recommendation algorithm, and as computers executing the designed algorithm. During the second phase, participants are grouped together. They are expected to present and argue about the merits of their respective algorithms and then more generally to discuss and reflect on their media practices.

During the first phase participants act as engineers. Each group receives a data sample about the videos to recommend, and need to devise a calculation rule general enough to enable them to rank the videos. The second step requires participants to act as algorithms. Each group receives, at regular intervals, new data that needs to be quickly processed, following the devised calculation rule, to produce actual recommendations. The participants are given limited instruction on how to proceed and are encouraged to collectively address possibilities and obstacles.

During the second phase, the teams discuss and evaluate their algorithms. Participants first present their own algorithm explaining their respective calculation rules and, finally, the groups vote on the most convincing recommendation algorithm. Then the media educator steps in, acting as a Chief Technology Officer, offering participants to evaluate, for them, their respective algorithms as most digital platforms would. Finally, the participants more broadly discuss the challenges encountered in the course of the activity, their own experiences of digital platforms, and the importance of recommendation in forming cultural habits.

Analysing the educational outcomes of the game

"In the shoes of an algorithm" essentially rests on an experiential learning process based on the participants' experience (knowing how) rather than an explicit and theoretical learning process (knowing that) based on the professors' knowledge.

As noted, the educational activity aims to enable participants to better think and talk about recommender systems (Gray et al., 2018). Learning may be thought to take place each time tensions between the participants' experience and expectation may be observed (Kolb, 1984). The triggers, indicating such tensions, may be (1) resorting to one's own experience (e.g., an intimate or a second-hand knowledge of the platform) to address a broader issue, (2) encountering an *obstacle* (e.g., a discrepancy between expected results and those produced by the algorithm) that propels an invention, (3) vivid argument about the decisions that are to be taken (e.g., from the more relevant data to the more desirable objective). These triggers acted as markers of where and when learning might take place.

The data used in our analysis essentially consists of partial transcriptions of two sessions from the nine sessions we attended. Each session comprised: one hour per group designing and executing an algorithm; and, for the second phase, one hour where all the groups discussed together. The first session (ESC) gathered students aged sixteen years old from the same class of a Belgian secondary school. The second session (CDS) gathered students aged twenty years old enrolled in the second year at a Belgian University. The remaining sessions comprised further sessions with secondary school students, university students and with university researchers and administrative staff. The following section analyses long and continuous extract of the sessions so as to render the participants' critical reasonings.

Results

During the first phase of the activity, small groups were invited to get acquainted with the data they received, to devise their own ranking procedures, and to process the provided data. The educators went from group to group: complementing the instructions received, helping them to overcome difficulties or, sometimes, challenging them. The interactions between participants often took the form of questions regarding the relevance of the data (e.g., I believe *this* number means *that*), arguments about the design of their ranking algorithm (e.g., Should we not change *this* weight?) and exchanges related to actual computations (e.g., Tell me the value of *this* cell). A few interactions also supported casual discussions about musical taste (e.g., I don't like *this* music!) or each other (e.g., I've known him since kindergarten!). The next four sections systematically analyse the interactions surrounding data, algorithms and computations.

The second phase of the activity consisted of one large group debating the relative merits of the various data and algorithms presented. It discussed recommendation processes and personal tastes, argued about the issues raised by recommender systems and digital platforms. The educators here led the discussion through rather open questions (e.g., Can you think about examples of other recommender systems?), or picking up on a question or a comment expressed by a participant (e.g., Do you agree with *this* idea?). The discussions, taking place in a larger group, adopted a more formal tone in which participants would account for the design choices they made, argue about the relative merits of different data and algorithms, or exchange their views of digital platforms. The final three subsections here analyse the interactions about algorithms and the digital platforms.

Grasping data

Participants begin by discussing the *data* they received. Analysis of the sessions uncovered three broad gestures traditionally associated with data literacy. First, participants sought to *contextualise* the data at hand, by considering its limits. Second, participants attempted to collectively *interpret* the data, by discussing the charts' readings and relating them to their experience. Third, participants sought to *extrapolate*, from the sample they received, to other data they might come across. The following extract usefully condenses these otherwise sparse forms of reasoning. The discussion here took place between participants discovering the data we provided to them: namely, demographic data about videos (to be ranked) and demographic data about a persona (for which they must be ranked).

First, participants contrast the data currently available (e.g., gender, age, location) with other data, assumed to be available (e.g., user history):

LISA: So, the elements composing the formula must be found in the data they give us?

ISMAEL: Yes, it seems.

LISA: Because her browsing history, for example, might also be very useful. Let us imagine she watches influencers' videos, but her twin sister watches wrestling videos, the suggestions won't be the same, but the formula must remain the same.

[ESC, User Data, 1/3]

Secondly, participants discuss what should the correct reading of the chart be, invoking their personal relation to the data presented.

YAN: Here they give us an example: "J'ai tout oublié" by Angèle is listened [to] more often by men, living in France, aged between 13 and 17.

LISA: Really?

ISMAEL: No, by women.
FANNY: Hmm, so Yan does not know how to read a chart.
LUKE: Indeed, it is more often listened to by women living in France.

[ESC, User Data, 2/3]

Third, a participant bluntly rejects an interpretative use of the personal relation one might have to the data.

YAN: Whatever, I listen to Angèle, and last time I checked, I was a man.
LISA: And? Is the chart based on you?"

[ESC, User Data, 3/3]

Interpreting data

Participants then further explored the significance of the data, interrogating the social processes from which data might have been generated; and secondly, considering the relevance of the data for particular tasks. As well as instantiating the two relations one can develop to data (data *from* processes and data *for* tasks), the following extract shows how key collective discussions are to a robust interpretation of data. This discussion took place some time after they have received their data dealing with statistics about videos (views, likes, comments, etc.) and channels (subscription, progression, etc.).

First, participants questioned the relevance of some data (e.g., comments, subscribers, views) by questioning their possible meanings (i.e., comments may be considered as positive or negative interactions, or yet as mere reactions).

NILS: Guys, we have forgotten the comments.
FRANCES: Yes, but that's a choice. Personally, I find comments obsolete.
MOUSSA: Yes, one can have 450,000 positive comments or 450,000 negative comments.
TIM: Nonetheless, it shows the popularity, because if people react, it means that it is relevant.

[CDS, Video Data, 1/3]

Participants discussed the relevance of data in relation to their still undefined objectives, aligned with their knowledge about how YouTube functions (i.e., rewarding videos with comments) or with other possible ways of functioning.

MONA: Actually, one earns more money if there are more comments.
FRANCES: After all, we are not forced to make like YouTube [...] we are not forced to copy YouTube's policy.
JERRY: Shall we vote on who wants to take comments into account? For me they have no value.

[CDS, Video Data, 2/3]

The discussion shows an essential role of argument in the process of forging a robust interpretation of the data at hand (e.g., data may be meaningless, evidence for positive interactions, information about people's attention)

FRANCES: The number of views could do, but the number of comments? anybody can say anything.
TIM: It shows that followers are engaged and active.
MONA: It shows that the video has caught people's attention, positively or negatively.
NILS: So, shall we take them into account?"

[CDS, Video Data, 3/3]

Designing algorithms

Participants had to invent a rule for ranking the videos received. They envisioned which kinds of operations might allow them to pass from data to rankings. Participants usually encountered two kinds of obstacles. Firstly, they might struggle to agree on what should be considered when recommending videos (What to discard? How to weigh? etc.). Secondly, they might have difficulties stabilising the operations that should produce recommendations – note that the structures of their algorithm may significantly vary with the kinds of data groups received (e.g., one-line formula, hierarchical trees, vernacular procedure). In the following extract, participants manipulating demographic data, assisted by the educator, overcome difficulties to agree upon what should be taken into account as it was turned into a stable series of operations.

After a struggle with the diversity of possible algorithmic design and the necessity of stopping one calculation rule participants were asked about the relative importance of the kinds of data (here: gender, age, and location) and about the objective of the recommender systems (here: music or tutorials).

EDUCATOR: So, do you take gender into consideration? Not location?
EMILY: We only take the age in consideration. Gender is not relevant for music. Although, it might well be for tutorials…
CARL: And the location is useless.
HIND: How do you deal with Chinese videos that have millions of views, then?
CARL: Maybe some people watch them, location is not relevant.
EDUCATOR: It is up to you. You need to understand that there are no good or bad answers. It is up to you to decide the formula that you find the most interesting.
EMILY: So for instance, the percentage associated with the age (13–17 years old, in the case Juliette) more the percentage associated with the location (here: Belgium).
EDUCATOR: That is one way of doing it.

[ECS, Demographic Data, 1/2]

The discussion shows participants hesitating as to whether the envisioned procedure is general enough so that it can be used to recommend any video (here: Angèle) to any profile (here: Juliette) – a concern pervasive in computing practices.

ARNOLD: But how to do that in mathematics?

EMILY: What?

ARNOLD: You cannot do it in English.

EMILY: In which language do you want to do it, then?

ARNOLD: In mathematical language. Here you only calculate one thing. We need to make a global procedure that could work on all kinds of songs.

EDUCATOR: And this is not good?

ARNOLD: It only works for 'Angèle'.

EMILY: One just has to replace.

EDUCATOR: She wrote in English 'Each time we have a video, we add the percentage of people that have the same age and the percentage of people that live in the same place as Juliette'... that is a way of formalizing, one can apply it to any video.

[ECS, Demographic Data, 2/2]

Executing computations

Finally, we invited participants to act as algorithms, having to process the data they received every five minutes or so, according to the algorithm they had just designed. The participants may face unforeseen cases, having to adjust the computational rules (e.g., when a tie happens between videos). They may be disappointed by their rankings (e.g., "The Beatles with the lower score, it breaks my heart") or realise the algorithm they designed happens to be too demanding in terms of computational time. In some cases, the participants even divided the computational labour – this requires a rule that can easily be broken down and supposes an equal involvement among participants. Hence, algorithms appeared not so much as formal disembodied entities, but as concrete operational processes, whose executions and results generally need to be assessed by trial and error.

Presenting algorithms

The general discussion held at the end of the game systematically started with a quick round, during which teams presented to each other, the data they had received and the algorithm they designed. The presentation demanded participants explicitly articulate a variety of dimensions: contrasting the material they received, justifying the data they considered relevant, explaining the rules and objectives their algorithm embedded, and summarising any computational difficulties. In the following extract, one student presents the algorithm his group designed, after having introduced the data they were provided (views, likes, subscribers, etc.):

TOM: First, we took the 'number of views of the channel' and we divided it by the 'number of views of the video'. We obtained a number and divided it by 10, to have a score on 10 [...] Then we took the 'likes of the video' and we subtracted the 'dislikes' from the 'likes'. It gave us an average, roughly, of what people liked [...] it allowed us to see whether or not a video was liked or disliked by YouTube users. Then we took the 'number of views of the video' and the age of the video and we divided the 'number of views of the video' by the 'number of months of the videos', so that we got an average of the number of views per month. And after, each time, we took the averages on 10 of the three computations I just described, and we did an average of the three, to rank the videos.

[CDS, Discussion 1/2]

The public exercise required them to produce an intelligible description of the algorithm's functioning, and to provide reasons that might ground the choices they made during the engineering phase (e.g., relating bits of the formula to social processes). They also had to recognise imperfections that might hamper their algorithms (e.g., the gap between aims and achievement). The teams' performances inevitably varied. The following passage, taken from a group manipulating demographic data (e.g., gender, age, location), shows one example of the difficulties sometimes encountered when accounting for calculation rules:

MARY: I will explain little by little... What we did is that we started from 15... and we always multiply 15 by 0, ½, 1 depending on the location, if it is respected. So, for the location we gave: 15 times 0 if it was 'Other' the higher, 15 times ½ if it was countries speaking English and 1 if it was countries speaking French, so we started with 15 times 0, ½ or 1.

EDUCATOR: So, you rewarded the video when the video was often watched in the same geographical zone as the person.

MARY: Yes, that is it.

[CDS, Discussion, 2/2]

Evaluating algorithms

The educators then invited participants to contrast the algorithms designed by the different teams and to then rank the algorithms they found the most relevant. The discussions generally focused on the kinds of data provided (and the power of generalisation presumably associated with them) – and less, on specific algorithms that had been presented. Participants often changed their positions in the course of the discussion (e.g., it would often take some time before participants realised the interest of viewing histories for recommending). In the debate transcribed below, participants argue about the capacities of data and algorithms while discussing their representations of users' preferences or engineers' objectives. They first discuss the relative merits of different recommender systems: arguing about the relevance of the data or the tension between personal and impersonal recommendations.

EDUCATOR: Does someone have opinions about which algorithms are more or less important?

OLIVER: They are all important.

AMELIA: Some focus too narrowly on specific users, so they might not be general enough.

NOAH: For me it is the opposite, some people like and dislike videos without really thinking about it: there may be haters or fake accounts. We could base our recommender system on the age of people having actually watched the videos, it remains general enough.

ZOE: But the problem with the data on age (and all) is that it is based on past data, of people having watched the videos, while here we have the rest of the data (likes, dislikes, etc.).

[ESC, Discussion 1/3]

Participants then discussed the various objectives the recommender system may need to consider. They mobilise, in the course of the argument, their own preferences, representations of users' wants or assumptions about YouTube's goals.

JACK: Personally, when I am on YouTube, the videos I want to see are videos close to the videos I usually watch and, sure, to have some novelty, but not too much.

MIA: This is rather subjective. Some prefer targeted videos. Others, not.

[ESC, Discussion 2/3]

These participants, we contend, gradually learn to think more like engineers, and to be able to critically reflect about the social consequences of the choices they made, sometimes even proposing alternative recommender systems to overcome some of the limits perceived (here: the possibility to choose the configuration of the recommender systems).

MIA: Maybe we should allow both.

EDUCATOR: So, users could choose?

MIA: Yes. [...]

HENRY: It also depends on if we want to please the public or the company.

EDUCATOR: How?

HENRY: If we give what the public wants, we will necessarily receive more money than if we give them something they may not want to watch.

[ESC, Discussion 3/3]

The educator would then close this part of the debate by asking participants to vote for the algorithms they find the most relevant and by placing an insert on how YouTube's Chief Technical Officer might have evaluated their algorithms – unpacking an intuitive case of metrics.

Discussing recommendations

The discussion process would then allow participants to recount their own experience and discuss wider societal challenges. All participants related mixed experiences about YouTube's recommender system, and they appeared keen to blame recommender systems. For example, recommendations were charged for being off mark by a student: "sometimes, I seek to watch similar videos and it just won't come up with anything interesting" [ESC, Discussion]. Another student accused them of locking him in: "for instance, if one listens to a rap song, it will only propose rap songs, but we might well wish to listen to something else: it locks us into a musical genre we already have" [CDS, Discussion]. They insisted on the interest of recommender systems that enabled the discovery of songs – and how they relied on them. Some participants also seemed to place conflicting expectations on recommender systems: "I want something out of the ordinary, not necessarily something known, not necessarily popular, that I am sure I will like because I behave like a lemming, I also want to get less well-known songs" [CDS, Discussion].

The participants sometimes succeed in moving from personal frictions to more general issues raised by recommender systems. Thus, the very personal sensation of being enclosed in a bubble is sometimes turned into a broader political argument about recommender systems producing uniformity: "We will shape people as we want them to be… we will lead everybody to the same things… everybody will like the same things… eventually, people will have less personality" [ESC, Discussion]. The systems were accused of perpetuating the established order:

> Before saying criteria such as gender, age, etc. are important, one must first see the cultural and social contexts… and then: do we decide to perpetuate a gendered culture? Or do we decide, with our algorithm, to break up these habits.
>
> [CDS, Discussion]

The emerging ability of participants to move from personal experience to more general accounts of recommender systems provided the basis for a collective and critical discussion.

Criticising platforms

The discussion finally brought participants to situate recommender systems within digital platforms and to imagine alternative ways of conceiving of them. The participants appeared to resort to three main arguments for better situating recommender systems within digital platforms. First, they sometimes proposed to articulate their experience being captured by the recommender system's objectives: "The objective is that we stay on YouTube […] their objective is that it pleases us, so we are willing to watch" [CDS, Discussion]. Secondly, participants often sought to articulate their second-hand knowledge about influencers' money-making practices in terms of assumptions about the platforms' strategy: "I have seen an influencer saying it: the longer the video is, the more advertisement can be displayed, and the more money can be made" [ESC, Discussion]. Third, they sometimes attempted to relate processes across several platforms: "When we do search on Google, we have suggestions,

I wonder: Is that an algorithm? Thus, it is not uniquely relevant what we see on YouTube, but also whatever we see on other websites." [CDS, Discussion] Hence, participants shared and enriched their knowledge of the dynamics in contemporary digital platforms.

The participants systematically sought to conceive alternative recommender systems. They often wanted to take into account features, subtler than the ones available to them or to YouTube's engineers: "It would be interesting to develop a programme that would analyse the comments" [ESC, Discussion]. They also imagined platforms that could let people choose between the different systems: "Ideally, we should be able to choose the kind of recommendation we want, whether we want to have recommendations that are diverse or similar to our tastes" [CDS, Discussion]. They proposed, although more rarely, recommender systems that could work on different grounds: "The movie critic recommends based on her expertise, the critic says, without knowing us, I find this movie is good for this and that reason… while YouTube does this on the basis of probabilities" [CDS, Discussion]. The educational activity appears to have stimulated an appetite for alternative recommender systems.

Discussion

The following sections respectively address two main questions. Firstly: What did the participants effectively learn? Secondly: What limits and perspectives does this educational activity raise?

Learning outcomes

The analysis shows that our educational activity gave participants an opportunity to improve their capacities to think and talk about recommender systems. From a technical perspective, the educational activity would generally open up with participants' fuzzy answers to the educators' opening question "What is an algorithm?" Nevertheless, the activity required participants to invent a sound way of turning a set of data about videos into a list ranking the videos, and seemed to have invariably brought them to articulate elementary technical capacities, at the core of algorithmic practices, as described in the first half of the preceding section: manipulating personal data, formalising ranking operations and dividing computational labour. The fictional situation inviting participants to act, in turn, as YouTube engineers and algorithms, enabled participants to acquire an understanding of algorithms as creations of situated engineers, rather than disembodied technical processes. Thus, participants were gradually brought to apprehend the social dimensions of these technical operations. The educational activity made recommender systems collectively discussible.

The activity and the exchange proved instrumental in bringing participants to discuss recommender systems critically. From a critical perspective, the educational activity allowed us to document three of the many sense-making strategies deployed by participants towards recommender systems. Firstly, through an enlargement of perspectives: moving from particular experience to more general considerations (e.g., from influencers' monetisation to YouTube's policies). Secondly, as the statement of a concern: turning a social experience into a political injustice (e.g., from pleasing recommendations to establishing order). And finally, with the imagination of other propositions: from the actual system to an alternative one (e.g., from an imposed to a chosen environment). The educational activity, we thus suggest, does indeed foster critical engagement with recommender systems in particular and digital platforms in general.

Our analysis shows how participants were brought to approach data and algorithms as design engineers as well as everyday consumers. This enabled participants to articulate technical constraints (e.g., data availability, mathematical formalisation, calculation complexity) and social objectives (e.g., advertisement money, user engagement, recommendation diversity). The activity not only provides participants with the means to understand how data and algorithms might be involved in recommender systems but, more significantly we hope, left them with an understanding of how, through metrics, engineering practices embed social objectives into socio-technical systems. Thus, the educational activity appears to have succeeded in treating the technical competences, not so much as *ends* for the activity, but rather as *means* for enabling participants to exert critical competence.

Limits and perspectives

The emphasis of Fastrez (2010) on the social, informal and technical competences characterising media literacy enabled us to more finely place the outcomes of our educational activity. While the activity undoubtedly fostered the articulation of the technical and social dimensions of recomender systems, it left relatively untouched the informational dimension (e.g., the interface of recommender systems or the effects of platforms' algorithms on their contents). The technical dimensions of recommender systems had to be limited, leaving important aspects outside the scope of the educational activity (e.g., the processing capacities and the learning algorithms are patently absent). The discussion of the social dimensions may have been limited by the very nature of the informational content staged during the activity. Music videos, ensure participants' early engagement, but unlike news, for example, may well hamper a more substantial discussion about recommender system consequences.

The educational activity wagered that the learning would prove more effective by immersing participants in a game-like simulation. The educators, acting as YouTube's Chief Technology Officers, invited participants to act, in turn, as engineers and as algorithms. It incorporated a plausible role-play dimension by creating a playful ambiance, using time limited elements in the activity to create a rhythm and by designating a winner in the classroom. The engagement of participants remained, overall, a success, although it strongly varied with group dynamics. However, the immersive character of the activity appeared to sometimes neglect the passage from doing to saying (e.g., skilful students would often publicly present). Further work might uncover neater ways of articulating the game-like activity to more formal educational setups.

Thus, for instance, the interpretive method used for assessing the educational outcomes of the activity, which largely stemmed from our critical stance and our experiential approach, cannot easily be articulated to more traditional forms of assessments used in formal education. However, the results here presented might be read as providing indications as to how formal education could take advantage of the informal activity, namely: by proceeding to a more traditional lecture (and evaluation) on data and algorithms and by designing a collective or individual exercise allowing professors to evaluate participants' arguments. These developments, on which we are currently working, shall enable us to more tightly relate the informal educational activity to more formal educational stakes.

Conclusion

We aimed to design an activity that would help us address the tension, experienced by many people, between a digital platform's significance and its opaqueness (the "black box"). The educational activity sought to improve participants' capacities to think and talk about recommender systems – also, providing them with the technical capabilities to engage in discussion about these matters. The game-like simulation confronted participants with an open problem impelling them to come up with a rule for passing from a set of personal data about interactions between YouTube's users and videos (inputs) to a list of ranked videos that might stand as YouTube's recommendation (outputs). They were brought to act, in turn, as engineers designing the algorithm and as computers executing the designed algorithm.

This chapter has addressed the effectiveness of our intervention by analysing recorded sessions from the educational activity. The analysis looked for triggers where the technical and critical enter the discourse. During its first phase, the activity brought participants to better grasp the concept of data with increasing precision, and to articulate computational rules and recommendation objectives, to better envision the ranking effects of their computational rules. During its second phase, the activity allowed participants to move from a specific to a general experience, to assess the fairness of a computational rule and to imagine alternative ways of recommending. The educational activity, we believe to have shown, succeeded in creating a space in which participants become capable of critically reflecting upon recommender systems as well as digital platforms.

We further hope the activity and our analysis of the results may inform other media education initiatives that take digital infrastructures as their object. In terms of content, the educational activity sought to set a fruitful balance between enabling participants to make sense of rather complex technical infrastructures, while leaving enough room to critically discuss how such infrastructures affect their media practice. In terms of method, the simulation invited them to penetrate inside the algorithms' manufacture, bringing them to approach otherwise abstruse technical devices as concrete processes mingling social and technical dimensions (e.g., filters, business, tastes). The experiential approach enabled participants to articulate their concrete media practices in more abstract discussions. This balance and our experiential stance appeared to be key in designing an educational initiative that fostered technical and critical competences.

Note

1 We prefer this term as used by the engineering community (see, for example the ACM conference) as opposed to the more common phrase "recommendation system".

References

Anderson, J. A., & Meyer, T. P. (1988). *Mediated Communication: A Social Action Perspective*. SAGE Publications.

Anderson, T., & Shattuck, J. (2012). Design-based research: A decade of progress in education research? *Educational Researcher, 41*, 16–25.

Aufderheide, P. (1993). *Media Literacy: A Report of the National Leadership Conference on Media Literacy*. Aspen Institute, Communications and Society Program.

Barab, S., & Squire, K. (2004). Design-based research: Putting a stake in the ground. *Journal of the Learning Sciences, 13*(1), 1–14.

Bijker, W. E., Hughes, T. P., Pinch, T. (Eds.). (1987). *The Social Construction of Technological Systems: New Directions in the Sociology and History of Technology*. MIT Press.

Buckingham, D. (2019). *The Media Education Manifesto*. Polity Press.

Claes, A., & Philippette, T. (2020). Defining a critical data literacy for recommender systems: A media-grounded approach. *Journal of Media Literacy Education, 12*(3), 17–29.

Cobb, P., Confrey, J., diSessa, A., Lehrer, R., & Schauble, L. (2003). Design experiments in educational research. *Educational Researcher, 32*(1), 9–13.

Desrosières, A. (1992). *La Politique Des Grands Nombres: Histoire De La Raison Statistique*, La découverte.

Fastrez, P. (2010). Quelles compétences le concept de littératie médiatique englobe-t-il? Une proposition de définition matricielle. *Recherches en Communication, 33*, 35–52.

Fastrez, P., & De Smedt, T. (2012). Une description matricielle des compétences en littératie médiatique. In M. Lebrun-Brossard, N. Lacelle, & J.-F. Boutin (Eds.), *La littératie médiatique multimodale. De nouvelles approches en lecture-écriture à l'école et hors de l'école* (pp. 125–140). Presses de l'Université du Québec.

Gitelman, L. (Ed.) (2013). *Raw Data Is An Oxymoron*, MIT Press.

Grosman, J., & Reigeluth, T. (2019). Perspectives on algorithmic normativities: Engineers, objects, activities. *Big Data & Society* doi:10.1177/2053951719858742

Gray, J., Gerlitz, C., & Bounegru, L. (2018). Data infrastructure literacy. *Big Data & Society, 5*(2), 1–13.

Hardy, L., Dixon, C., & Hsi, S. (2019). From data collectors to data producers: Shifting students' relationship to data. *Journal of the Learning Sciences, 29*(1), 1–23.

Hill, R. K. (2016). What an algorithm is? *Philosophy and Technology, 29*(1), 36–59.

Hobbs, R. (2010). *Digital and Media Literacy: A Plan of Action (Communications and Society Program)*, p. 67. The Aspen Institute.

Jacques, J., Fastrez, P., & De Smedt, T. (2013). Organising media as social objects: An exploratory assessment of a core media literacy competence. *The Media Education Research Journal*, *4*(1), 42–57.

Jacques, J., Grosman, J., Collard, A.-S., Oh, Y., Kim, A., & Jeong, H.-S. (2020). "In the shoes of an algorithm". a media education game to address issues related to recommendation algorithms. *The Journal of Education*, *3*(1), 37–62.

Jacobi, D. (2001). Savoirs non formels ou apprentissages implicites?. *Recherches en Communication*, *15*, 169–184.

Jaton, F. (2017). "We get the algorithms of our ground truths: Designing referential databases in digital image processing." *Social Studies of Science*, *46*(7), 811–840.

Jørgensen, M. W., & Phillips, L. J. (2002). *Discourse Analysis as Theory and Method*. SAGE.

Kolb, D. (1984). *Experiential Learning: Experience as the Source of Learning*. Prentice-Hall.

Ligurgo, V., Fastrez, P., Jacques, J., Collard, A.-S., & Philippette, T. (2019). A definition of digital media literacy competences required by workers to collaborate in distance work environments. In J. Jacques & A.-S. Collard (Eds.), *Digital Media Literacy in Teamwork and Distance Work: Competences, Discourse and Organizational Design* (pp.13–68). Namur.

McKenney, S., & Reeves, T. C. (2013). Systematic review of design-based research progress: Is a little knowledge a dangerous thing? *Educational Researcher*, *42*(2), 97–100.

Pangrazio, L., & Sefton-Green, J. (2020). The social utility of 'data literacy'. *Learning, Media and Technology*, *45*(2), 208-229. doi:10.1080/17439884.2020.1707223

Porter, T. (1996). *Trust in Numbers: The Pursuit of Objectivity in Science and Public Life*, Princeton University Press.

Rancière, J. (1983). *Le Maître Ignorant: Cinq Leçons Sur L'émancipation Intellectuelle*, Fayard.

Ricci, F., Rokach, L., & Shapira, B. (Eds.). (2015). *Recommender Systems Handbook*, Springer.

Saariketo, M. (2014). Imagining alternative agency in techno-society: Outlining the basis of critical technology education. In L. Kramp, N. Carpentier, A. Hepp, I. Tomanić Trivundža, H. Nieminen, R. Kunelius, T. Olsson, E. Sundin, & K. Kilborn (Eds.), *Media Practice and Everyday Agency in Europe* (pp. 129–138). ed. lumière.

Seavers, N. (2019). Captivating algorithms: Recommender systems as traps. *Journal of Material Culture*, *24*(4), 421–436.

Selwyn, N., & Pangrazio, L. (2018). Doing data differently? Developing personal data tactics and strategies amongst young mobile media users. *Big Data & Society*, *5*(1). doi: 10.1177/2053951718765021

Seymoens, T., Van Audenhove, L., Van den Broeck, W., & Mariën, I. (2020). Data literacy on the road: Setting up a large-scale data literacy initiative in the DataBuzz project. *Journal of Media Literacy Education*, *12*(3), 102–119.

Simondon, G. (1958). *Du Mode D'existence Des Objets Techniques*. Aubier.

Slayton, R. (2015). "Measuring risk: Computer security metrics, automation, and learning". *IEEE Annals of the History of Computing* *37*(2):32–45.

Star, S.L., & Ruhleder, K. (1996). Steps toward an ecology of infrastructure: Design and access for large information spaces. *Information Systems Research*, 7, 111–134.

The Design-Based Research Collective. (2003). Design-based research: An emerging paradigm for educational inquiry. *Educational Researcher*, *32*(1), 5–8.

Wang, F., & Hannafin, M. (2005). Design-based research and technology-enhanced learning environments. Educational Technology Research and Development. *Educational Technology Research and Development*, *53*, 5–23.

9

CRITICAL ALGORITHM LITERACY EDUCATION IN THE AGE OF DIGITAL PLATFORMS

Teaching children to understand YouTube recommendation algorithms

Hyeon-Seon Jeong, Yeon Ju Oh and Amie Kim

Introduction

With the rapid increase in the use of personalised digital media devices and social media worldwide, digital platforms have used algorithms to create segmented audiences from personal data. South Korea is no exception to the growing influence of digital platforms and their algorithmic power. YouTube is where South Koreans spend the most time, recording 46 billion minutes of viewing time during August 2019. The number of monthly YouTube users increased by 7%, from 30.93 million in August 2018 to 33.08 million in August 2019, accounting for 64.7% of the whole population. In particular, children appear to be heavy users of YouTube; 90.3% of elementary school students in South Korea use YouTube, and 43.5% use it every day (Bae et al., 2020).

YouTube users are shown a list of recommended videos when they enter the main page or watch a video, all of which are chosen by algorithms based on individual viewing records. This so-called recommendation system is used to further increase viewing time by suggesting videos that users are likely to watch. YouTube algorithmically mediates the viewing experience by collecting data whose meanings are only interpretable to experts skilled in reverse engineering or insiders. In this way, YouTube shapes our ideas and beliefs about culture, society and politics. As YouTube has become the most popular social media platform and is used as a source of news, education, entertainment and learning, it is important for wider social and apolitical purposes to interrogate how YouTube designers and developers mediate content through algorithmic practices.

This chapter responds to the pedagogical challenges faced by media educators how to unravel the hidden algorithmic practices in classroom settings. International scholars and educators in Belgium and South Korea closely collaborated for two years to develop a media literacy education programme designed for a critical

DOI: 10.4324/9781003136842-9

understanding of the recommendation algorithm of YouTube (see the previous chapter). This collaboration began when two authors of this chapter participated in the 2018 Media Education Summit in Hong Kong, where the Belgian researchers presented the development and implementation of a media education game, "In the Shoes of an Algorithm", for high school students. The Belgian game aimed to create an algorithm that would most appeal to the YouTube CEO. Students were provided with raw data on YouTube users as well as videos and asked them to analyse the dataset. We became interested in adapting this game for critical algorithm literacy education in the South Korean context.

This chapter discusses the South Korean adaption of the original version of the Belgian game, specifically the underlying pedagogy that provides guided activities for the learners to realise that YouTube's recommendation algorithm is value-laden and open to revisions and improvements. In contrast to the informal education described in the preceding chapter, the South Korean programme was designed for upper elementary students, considering that the age of children using digital platforms such as YouTube is becoming lower, and most upper elementary students are users of smartphones and YouTube (Jacques et al., 2020). The recent statistics show that 87.7% of the 4th to 6th graders aged 9 to 12 have their smartphones, and YouTube is the most frequently used application (Bae, Lee and Lee, 2020). In this context, media education focusing on a critical understanding of YouTube's recommendation algorithm seems to be salient for upper elementary students given their media consumption. Thus, in adapting the Belgian game, we tried to create a programme simple enough for those students to grasp how recommendation algorithms of digital platforms might work. The newly developed critical algorithm literacy programme (hereafter, CAL) provided learners with an opportunity to think as YouTube engineers by following the guidance of a teacher by exploring datasets of an imaginary YouTube user, and recommending lists of videos to the user.

The CAL programme was implemented six times to different groups, including upper elementary students, high school students, college students and media educators, with some revisions and variations customised to each group. This chapter focuses on the version applied to 5th grade elementary students aged 10 to 11 in a formally organised afterschool care programme. We discuss the pedagogic design of the programme, focusing on the role of the teacher as a guide in developing a critical understanding of YouTube's recommendation algorithm on the basis that the algorithm plays an important role in what children watch and learn. This chapter adopts a critical perspective on datafication as a new issue for media literacy education in the age of digital platforms.

Education about datafication in South Korea

In the South Korean context, educational responses to datafication comes from two primary perspectives: the protectionist and the functional. Education from the protectionist perspective aims to enhance the ability to protect the personal data both of the individual and others. Data protection education began in the early 2000s as a part of a state-led initiative to strengthen information and communication ethics, referring to the basic principles and behaviours regarding emerging phenomena in the information society. In the late 2010s, as the need for new civic consciousness emerged along with the rise of digital transformation, the ability to protect personal data started to be treated as an element of 'digital citizenship.' While information and communication ethics education focuses on individual consciousness and behaviours, digital citizenship education expands its scope to appreciate participation and praxis for wider communities (Choi, 2016). However, the content of education under both terms were not fundamentally different in that the protectionist perspective pervades digital citizenship education.

Education from the functional perspective underscores technical data skills such as data mining, analysis and visualisation. The emergence of big data in the early 2010s and the discourses surrounding the 4th industrial revolution and artificial intelligence in the mid-2010s triggered the need to cultivate human resources equipped with data competency. As a result, the Ministry of Education has gradually introduced coding curricula to the national education system since 2018 and plans to implement an AI curriculum by 2025 fully. As the national curriculum standards are applied without school- or local-level autonomy, all students in the public education system are expected to learn how to code as a part of the National Curriculum.

From the protectionist perspective, datafication is perceived as a crisis that negatively affects personal safety. However, it is also regarded as a national and individual opportunity that would boost economic growth and reduce unemployment from the functional perspective. This dichotomous framework of crisis versus opportunity is not just a response to datafication but is deeply rooted in the overall media education system in South Korea. Although crisis and opportunity seem to be at opposite poles, both positions share a commonality. They reinforce highly personalised and normative behaviour, whether self-protection or self-enhancement. Such self-disciplinary education is also in line with the neo-liberalisation processes in South Korea, which have been accelerated since the economic crisis in the late 1990s. On the one hand, individuals must self-care without adequate safeguards against accelerating data capitalism and surveillance. On the other hand, they are forced to be productive and improved based on datafication in the face of declining economic growth and crises in employment.

Theoretical framework

Buckingham (2019) calls for a more consistent and holistic media education, criticising such dichotomous thinking leading to oversimplified, fragmentary and contradictory educational responses. Resonating with Buckingham's arguments on media education and Couldry's critical view on datafication (Couldry, 2020), the CAL programme was designed with three foci to promote a holistic view. Datafication is a term coined by Kenneth Cukier, editor of *The Economist*, and Viktor Mayer-Schoenberger, professor at the Oxford Internet Institute, meaning "taking all aspects of life [and] turning them into data" (2013, p.35). The production of such data is inextricably linked with the elements of data extraction, such as data collection, processing, refinement and value creation, such as revenue generation, state monitoring and citizen participation from the extracted data.

The first aim in developing the programme is to develop understanding of data extraction. While the term datafication has recently been in wide circulation, Andrejevic pointed out back in 2007 the dangers of datafication (Andrejevic, 2007), especially issues of surveillance and control, which were obscured by the potential of democratic use of interactive media. As academics and journalists have actively raised issues related to collecting and using personal data by businesses and governments, individuals' awareness of data has increased. According to a survey conducted in the USA by the Pew Research Center in 2019 (Auxier et al. 2019), the majority of respondents were concerned about the use of data by businesses (79%) and governments (64%). However, these concerns are not based on an understanding of the operation of the infrastructure that enables datafication or the assumptions and prejudices that inevitably arise in categorising human behaviours and social phenomena into specific categories.

The second focus is a critical understanding of the social order that is re/arranged by datafication. The value produced through datafication affects the economic, political and social domains beyond the individual level. According to the previously cited Pew Research Center survey, the percentages of respondents who understand the use of data collected by businesses and governments were relatively low at 40% and 21%, respectively. A survey by the Korea Communications Commission also showed that the percentage of respondents who were aware of their data being collected was only 34% (Bae et al., 2011). The legal definition of personal information in Korea is so comprehensive that there is insufficient social consent for the scope and type of personal information. According to a more recent study that surveyed users' perceptions of 18 personal information items, about 95% of adult respondents in their 20s and 50s responded that mobile phone numbers, social security numbers and financial information would be regarded as personal information.

In contrast, only 64.5% of the respondents answered that public posts on social media would be categorised as personal information (Choi et al., 2020). As regulations and legal measures on data protection and ownership are emerging, such as the General Data Protection Regulation of the European Union, individual rights over data have been established as an educational focus. However, beyond the issues of protection and ownership, there is a need for discussion and deliberation on how datafication transforms people's lives and the surrounding environment and how individual agency can be diminished through the processes of datafication.

Lastly, an emphasis is put on understanding the workings of economic and political power that make datafication possible. It is gradually becoming common sense that the big platform companies such as Google, Apple, Facebook, and Amazon and small-scale websites and applications use personal data for profit generation by developing technological interfaces that make datafication easier. Through the example of countries such as China, awareness of the government's data surveillance is increasing, and the complicity between governments and private companies in collecting and utilising data has attracted academic attention. Starting from the realisation that there is economic and political power behind datafication, it is crucial to contemplate how the power dynamics play out and what interventions can be implemented.

We used the concept proposed by Gray et al. (2018) of data infrastructure literacy and van Dijck's (2020) platformisation tree metaphor as the main theoretical framework in our analysis. Gray et al. point out the limitations of existing data literacy education centred on numerical skills and propose data infrastructure literacy as an alternative. According to them, "data infrastructures are shifting relations of databases, software, standards, classification systems, procedures, committees, processes, coordinates, user interface components and many other elements which are involved in the making and use of data (p.3)." Considering this definition, data infrastructure literacy refers to the ability to critically explore the relationship between objects and non-objects involved in producing and using data. The CAL programme presented in this chapter helps to conceptualise the data infrastructure by letting students think about the relationship between objects and non-objects, such as criteria, weights, developers, subscribers, viewing time, keywords and thumbnails, all of which have tend to be thought about in fragments.

Van Dijck (2020) proposes "the platformisation tree", a visual metaphor to imagine holistic governance for platform companies. In this metaphor, the trunk signifies platform companies or intermediaries such as Facebook and Google, and the root indicates the hardware foundation that allows such companies to be embedded. Branches denotes sectoral applications offered or mediated by platform companies, and lastly, oxygen refers to the essential ingredient fed into applications to function and grow. As one of the most familiar applications to children, YouTube functions as a heuristic tool to ponder the intertwining relationships among hardware, platform, application and data. The CAL programme was designed for young children to view YouTube as a venue to watch videos and a complex ecosystem consisting of various technical and social components.

Developing a media education programme about recommendation algorithms for children

In South Korean elementary schools, 5th and 6th graders (aged 10 to 12) learn computer programming in the Practical Course curriculum, mainly focusing on the algorithm as a way of procedural thinking to solve given problems. Students might learn where and how software might be used in daily lives, why software might be significant, and have hands-on programming experience. For hands-on programming experience, children might be given a problem to solve, asked to divide it into pieces, and then set the order (i.e., to make a procedure, such as in a flow chart), to solve it step by step. The focus is on experiencing the thinking process (i.e., computational thinking that can be used according to the problem-solving procedure). Finally, children have experience creating a simple program using block-based programming languages such as MIT Media Lab's *Scratch* or Naver Connect Foundation's *Entry*. As the course underscores technical aspects of algorithms, it is hardly related to media literacy education.

Media literacy education aims to help students critically reflect on their use of media and understand the role and influence of media technology in their lives. With the rapid increase in the use of personalised digital media devices and social media, user data produced by individuals have contributed to creating algorithms aiming to personalise the content on digital platforms. Therefore, a critical understanding of digital technology in media literacy education needs to be extended to recommendation algorithms to examine the designers' socio-cultural values and economic intentions. "In the Shoes of an Algorithm", the game developed by Belgian researchers and practitioners, pursues such a goal by encouraging students to understand how the recommendation algorithms work and how they affect the encounters users have with YouTube (see previous chapter). We found that the Belgian game would have significant implications for developing concrete methods of media literacy education that enable children and adolescents to be aware of how digital platforms operate. In particular, we found the Belgian game an excellent defamiliarising strategy that gave students a fresh and alternative view of YouTube – one which is not apparent to the typical user.

In adapting the Belgian game, the priority was making the programme simpler so that upper elementary student could participate. Having students take the role of an engineer remains central to the adapted version. However, it seemed difficult for younger children to create an algorithm based on data analysis. Considering the level of mathematical competency of upper elementary students, drawing the criteria for recommending videos and creating a formula by weighting the criteria was also tricky. Therefore, it was decided that the South Korean version would focus on ranking the order of the videos to recommend to possible users by placing video cards on a board and rearranging them according to the rankings based on students' criteria and then debriefing the reasons for doing so.

We also decided to visualise the data in order for students to grasp elements involved in YouTube's recommendation algorithm and easily participate in the task of selecting criteria for recommendation. Therefore, we used a strategy of constructing two kinds of datasets. The first dataset is a "Persona". Persona is a concept introduced by Alan Cooper (1999), referring to a fictional character representing a specific group of users. The Persona in the CAL programme is a character that students may consider as their other selves. The second dataset is nine video cards with relevant information such as thumbnail, title, description, the number of views, likes, subscribers and upload time, which could be taken as the criteria for recommendation. While the Belgian game provides the participants with 20 music videos recommended to one male and one female user respectively, only one fictional user and nine video cards were presented in the CAL programme. We searched the latest popular YouTube videos and channels and asked elementary students about their preferred content to select YouTube videos. Video cards were made to provide students with information for the recommendation criteria to read the basic information presented on the card about the content and use it as criteria for a recommendation without having to play each video on a digital device.

A block of two 40-minute classes was planned, which is somewhat short, considering the context of the South Korean elementary school where the teacher's control over the school curriculum is quite limited, and there is no regular class time dedicated to media education. The CAL programme was piloted in the afterschool care classroom of the 5th grade aged 10 to 11. The programme consists of relatively more structured activities that elementary school students can quickly complete. However, the teacher was advised to minimise direct intervention as much as possible so that the students could share their experiences, discuss the information presented in the given data, and select criteria that can be considered when recommending the videos. As the previous chapter showed, the Belgian game consisted of relatively less structured activities to encourage students to ponder without the teacher's intervention.

Data sets: persona, video cards and the activities

In the pilot study, six female and two male students participated. They were given one Persona Card and nine YouTube video cards as described below.

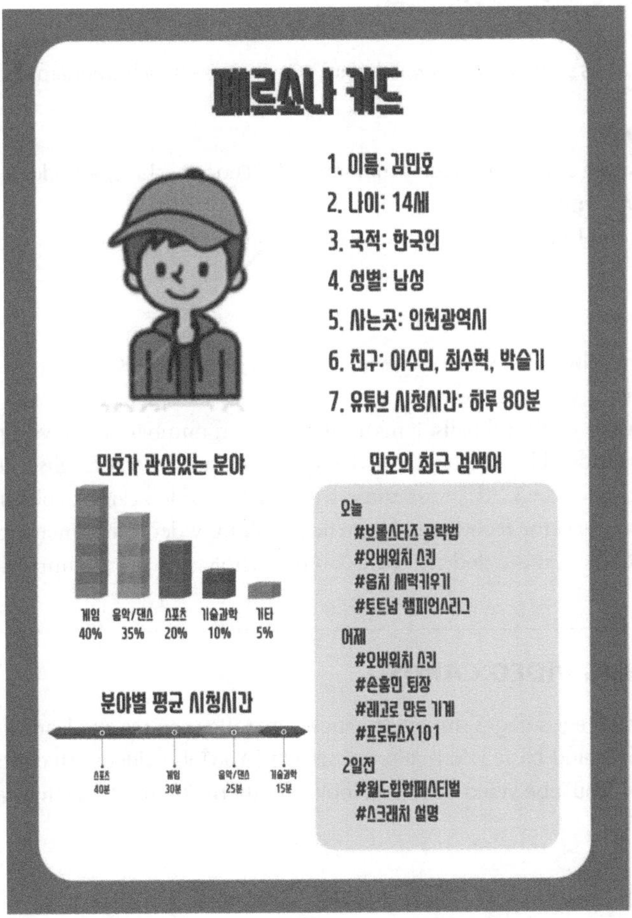

FIGURE 9.1 Persona Cards.

Name: Kim Minho
Age: 14 years old
Nationality: Korean
Gender: Male
Area of Residence: Incheon Metropolitan City
Friends: Sumin Lee, Suhyuk Choi, Seulgi Park
YouTube Viewing time: 80 minutes per day

Minho is Interested in:
Games 40%, Music/Dance 35%, Sports 20%,
Technical Science 10%, Others 5%
Viewing Time by Genre:
40 minutes for sports, 30 minutes for games, 25 minutes for music/dance, 15 minutes for technical science

Minho's Recent Search Terms
Today
Brawl Stars strategy, #Overwatch skin, #Opchi growth, #Tottenham Champions League,
Yesterday
#Overwatch skin, #Son Heung-min sent-off, #Lego machine, #Produce X 101
Two days ago
#World Hip-Hop Festival, #Scratch explanation

The YouTube videos recommended for the Persona, "Kim Minho", included various genres, whereas the Belgian game-used music videos only. Each video card was composed of thumbnails, length, title, creator, number of views, upload time, brief description. The languages in the video included both English and Korean. The activities for the CAL programme were reading video cards, exploring Persona, discussing criteria for recommendation and ranking video cards, meeting an imaginary YouTube engineer, and reflecting on the whole activity, as summarised below.

READING VIDEO CARDS

The teacher encourages students to know what they can see and learn about the thumbnails and basic information of selected YouTube videos. Students pulled out a few YouTube video cards and looked at them, interpreting information on the cards.

Exploring Persona

The teacher presents a specific Persona and presents the task (You should recommend a YouTube video to this Persona.) The students look at the Persona Card and analyse information.

Discussing Criteria for Recommendation and Ranking Video Cards

The teacher has the students display the video cards that they would recommend to the Persona through discussion.

Students talk about what videos they would like to recommend to the Persona. At this time, students share opinions with peers about information or criteria that should be considered necessary when recommending videos. Together, all participating students arrange YouTube video cards in the order of ranking of recommendations.

Meeting an Imaginary YouTube Engineer and Reflecting the Activities

The teacher invites all the students who participated in reviewing their activities and explaining how they negotiated the ranking of the recommendation criteria.

Assuming that the students have met an imaginary YouTube engineer, the teacher tells them what they thought was important when deciding the order to discuss the criteria for recommending videos.

Students share what they felt through these activities.

FIGURE 9.2 Activities of the Critical Algorithm Literacy Education Programme.

Findings from the pilot programme

In the process of adapting the Belgian game for upper elementary students in the South Korean context, researchers, teachers, and media activists closely collaborated to design the pedagogy through a series of structured activities in order to guide and facilitate exploring, questioning, listening, communicating and negotiating as group activities.

First, in discussing the criteria for selecting videos to recommend for the Persona, students showed their thoughts and reflections from various perspectives on the factors that influence their favourite YouTube content as YouTube users.

As seen in the Persona Card (see Figure 9.1 above), students were given information about a fictional character named Kim Minho. For example, his most recent search terms were online game-related terms (e.g., # Brawl Stars strategy, #Overwatch skin) and soccer-related ones (e.g., #Tottenham Champions League). When deciding which content should be recommended to Minho (the Persona), some students tend to choose content related to Minho's search terms., while others suggested recommending content that they, as teenagers, prefer.

In addition, in light of their YouTube viewing experience, the students carefully examined and discussed whether the length of the video was appropriate. When deciding which content to recommend to Minho, students suggested that 10 to 15 minutes is the appropriate amount of time to watch one clip on YouTube. They argued that if the content length is longer than 15 minutes, they tend not to watch the whole clip and rather skip it through. In this process, as YouTube users, we could get a glimpse into what the students judged to be the 'right' length.

Other than that, students considered whether the language of the content is in Korean or English and whether it is content that is not a topic of interest but maybe curious in the light of the Persona's search terms and viewing experience presented. As seen in students' discussion above, the way students approached the game has ample implications for media education. It suggests that media education dealing with algorithms or data literacy helps students understand how the YouTube recommendation algorithm works and allows students to discuss the criteria they value themselves as YouTube users.

Secondly, after going through the previous processes, students had the opportunity to think like an engineer who makes an algorithm for YouTube. The focus of this activity is to experience the way recommendation algorithms work on the digital platforms visited daily. Here, the CAL programme helped students think through how data works in recommending and arranging media content. As mentioned earlier, Korean elementary school students are learning algorithms focusing on computer logic and procedural thinking. We believe that in the process of students participating in this programme, the algorithm can be viewed as not just a procedure for operating the machine, but where humans give the final weight in the process of making the algorithm, and this problematises which criteria are critical. In other words, we tried to use a media education intervention to understand the importance of thinking about how to determine the weight of variables. In the process of playing the game and discussing the results, the students could see that social values are involved in determining and applying the algorithm to digital content. However, the students were confused about creating an algorithm that reflected their values – what they wanted – or creating an algorithm based on their YouTube experience. When we apply this game in the future, it is necessary to present one of these two goals and discuss them.

Discussion – towards data and platform imaginaries

The original game developed by the Belgium research team was aimed at older upper secondary school students. While adapting the game for upper elementary school students, the Korean version was designed as a 'simulation activity', often used in media education classes. Simulation activities are teaching and learning strategies that mainly use role-playing or games aimed at experiential learning. In media education, teachers can guide students to walk in the footsteps of real-world media professionals. They can understand how media content is produced and circulated as products and how media professionals work as a team within the media industry. Media educators use simulation activities to offer students opportunities to learn media production skills and encourage students to think about how media products are planned, produced and circulated. Through the simulation activity, students can understand the socioeconomic context of media production and consumption.

As we consider it essential for school education to allow students to link their daily lives to the official school curriculum, we decided to choose this kind of simulation activity instead of directly transferring knowledge about the social media algorithm to students. However, when analysing the classes, we recognised some unexpected gaps between how the activity was designed and how students participated. In addition, we realised that due to the changing media environment, it is essential to think which further steps should be taken to extend media literacy education into teaching and learning that can support students' critical algorithm literacy. For instance, to think and act like YouTube's algorithm developers, students need to think of the data they can collect and extract from users and analyse and make use of it to suit the company's business purpose.

This section suggests ways in which the game can be further developed to help students think critically about algorithms, especially how media literacy education 'simulation' activities can become a starting point for critical algorithm literacy. Firstly, as the 'simulation' activity encourages students to think and act as real-world professionals, teachers need to give clear guidance on what being in the footsteps of real-world producers means for the simulation activity. This is especially important when dealing with social media whose production methods (e.g., how the algorithm is planned and works, what types of information are collected and how they are used) are not transparent to outsiders.

Analysing the classroom activities, we recognised that some students went through the simulation activity trying to think as the real developer working for

YouTube during the game. At the same time, some of them positioned themselves as developers who were creating alternative algorithms that can be more user-friendly to their peers (and the Persona in the game). For instance, some of them mentioned that the recommendation algorithm can consider the user's preference or interests which can be inferred from the user's key search terms. As mentioned in the previous section, students' prior experience as active YouTube users intervened during the activities. Most of them had experience using YouTube for more than two years, which led them to gain an experiential understanding of how YouTube's recommendation algorithm works. Their personal experiences on YouTube had provided them with opportunities to think of preferred ways of how the recommendation algorithm should work. Therefore, teachers need to make it clear to students what aims simulate the professional's experiences. The programme can encourage students to consider the social values embedded in YouTube's recommendation algorithm and encourage them to develop an alternative recommendation algorithm for YouTube based on the reflection and discussion of their prior experiences.

Secondly, the simulation activity should lead to further discussion of the social meanings of the recommendation algorithms. The game we developed included reflection and discussion activities with students and the teacher to think of the social meaning of the recommendation algorithm and what it means for them. However, due to the limited amount of time, the reflection part was omitted. If this game is to be applied in other schools, we think it is necessary to wrap up the game to reflect the algorithm's social roles. They can think about the algorithms' implications in the short and long term. For instance, they can reflect on how social media's algorithm influenced their choice of content or how they communicate via social media. They can have further discussions about living in the digital society where this type of recommendation algorithms influences our daily lives. Furthermore, the class can discuss alternative ways for social media and recommendation algorithms to work, leading to speaking up and being an active part of digital governance. Linking this to the first point, the teacher can also design the simulation activity to encourage students to develop an algorithm that can avoid the formation of a 'filter bubble' or help users broaden their interest and communication.

Thirdly, the game could be further developed to help students become literate in the current media environment, where learning how to read multiple digital texts and data is necessary. Observing the classroom, we recognised that students were getting used to probing the information they acquire on YouTube using the YouTube card (see Figure 9.2). Students noticed that YouTube offers diverse information about uploaded content. It offers information about length, primary language, date of production, view count, numbers of likes and dislikes it acquired, etc. Students also discussed the visual image provided in the format of thumbnail pictures and subtitles. In addition, as mentioned before, the card used in the game followed the model of a computer or PC-based YouTube. Whether it is the same format students encounter when using YouTube can be probed in the class, as different devices and operating systems present YouTube differently. Finally, teachers can encourage students to think about what the 'digital divide' can mean in today's media environment and how it can influence our daily lives as digital citizens.

As discussed above, we would like to suggest that media literacy education in digital platforms should not be limited to understanding current YouTube algorithms but should encourage students to pay attention to the 'data' they produce and how the media industry extracts and uses it. For the media simulation activity illustrated in this chapter to become a starting point of critical algorithm literacy, we need to consider what students need to know about the datafied society and reach the learning goal when designing the learning experiences.

The CAL programme facilitates experiential learning by putting students in a position as a developer and helps them enact algorithmic and platform imaginaries. Algorithmic imaginary, a term coined by Bucher (2017), means "ways of thinking about what algorithms are, what they should be, how they function, and what these imaginations, in turn, make possible". According to Bucher, people encounter situations where they think, talk and feel about the algorithms behind relationships that appear online or in a particular application. "YouTube algorithms brought me here", a comment often added to YouTube videos highlight the moment when the algorithmic imaginaries are enacted.

Platform imaginary, a term conceptualised by van Es and Poell (2020, p.1), refers to "ways in which social actors understand and organize their activities about platform algorithms, interfaces, data infrastructures, moderation procedures, business models, user practices, and audiences". The emphasis of the CAL programme is the imaginaries presented by Bucher, van Es and Poell. These imaginaries help think about data infrastructure and platform ecosystems, to view algorithmic practices as value-laden, and imagine a better governance system of digital platforms in a data-driven society. Cultivating imaginative power is the very nature of critical algorithm literacy the CAL programme defines. Despite everyday encounters with YouTube and its algorithm, we do not concretely picture who is behind the algorithms we encounter, what data are used, how recommendations are made and how users are constructed if there is no particular moment that invokes imagination. The CAL programme invites students to imagine the inner workings of YouTube and its algorithm by simulating the processes of making sense of the data.

Platform imaginaries are important for a critical understanding of algorithms because algorithmic practices are increasingly becoming opaque. As machine learning, as well as human intervention, affects algorithmic practice, it is becoming opaque even to developers as well as users. When the CAL programme was introduced through conferences and papers, criticism was raised that the algorithm practiced in this programme is quite different from the actual YouTube recommendation. However, it is difficult to grasp the algorithmic practice of the YouTube recommendation system, which is neither clearly explained to developers nor transparently disclosed to the public. An understanding of the actual algorithm does not necessarily lead to critical thinking about datafication. The purpose of the CAL programme is not to understand exactly how the YouTube algorithm works but to help children understand the value-laden nature in which algorithms might be constructed. One of the core concepts of media literacy is that media is constructed, and thus it could be constructed in different ways (Buckingham, 2019). This core concept should be extended not only to content but also to platforms such as YouTube, of how algorithms could be constructed in different ways. We need to expand platform imaginaries to include critical thinking about YouTube.

Conclusion

One piece of feedback we often received during the planning and piloting of the CAL programme was whether it was within the realm of media education. Behind such feedback, there seems to be an assumption that media education fosters the ability to access, analyse, evaluate and produce texts or messages, that is, media literacy in the traditional sense. However, as platform companies are rapidly changing the media environment, media education focusing on the traditional concept of media literacy is limited. Digital platforms such as YouTube provide lists of personalised contents recommended by the algorithm, which mediates particular world representations. In the digital environment, critical analysis needs to be applied only to contents and platforms of media so that the alternatives of media construction could be imagined and sought. In a situation where platform companies are positioning themselves as technology companies, avoiding accountability for content production, mediation, and distribution and trying to avoid regulation, existing media literacy education fails to contribute to critically thinking about platform capitalism and develop the ability to intervene.

The CAL programme offers significant implications for media education at present when so-called technology companies are expanding their scale horizontally and vertically, creating new media environments. First, media education in the digital environment needs to enhance the perspective of the political economy to understand how content is produced, distributed and consumed beyond content-focused education. A critical understanding of media content is still essential. However, if media education is a critical understanding of media, it is also essential to explore the political and economic factors that affect the environment.

Next, it is necessary to expand the concept of media content. In video platforms such as YouTube or social networking services, media content usually refers to textual and visual products produced and uploaded by users. Scholars such as Napoli and Caplan (2017) argue that these companies are media companies, not mere technology companies, in that they have editorial power over content. In this chapter, we take this argument a step further. We suggest that these services, regardless of whether they are referred to as platforms or SNSs, should be seen as media in their own right in the sense that they mediate information, knowledge and data. Under this assumption, all the symbols that are represented in this services-in the case of YouTube, the personal information form that must be filled out when signing up for the service (thumbnails, upload dates, views, hashtags, video arrangements, etc.) enters into the realm of content to critically explore. In this way, digital data can be considered as a 'text' for 'data literacy' (Pangrazio and Sefton-Green, 2020) as a part of critical literacy. This expanded concept of media content can provide the starting points where children begin to unfamiliarise the familiar 'texts' in their media experiences.

Livingstone and Polizzi (2019) emphasise that media literacy should pay more attention to the role of digital technology and that it is crucial to understand how black-boxed automation systems might make differentiated choices. It is found that children thought that their privacy is essential. However, they tended to understand it in terms of interpersonal relationships rather than in the context of how their data might be used by Internet companies or public institutions (Livingstone et al., 2019). Children found it challenging to understand how Internet companies or public institutions might use their data. Parents might be worried without knowing how their children's data might be used. The age at which regular YouTube can be used is 13 years old. However, it was found that 32.6% of YouTube users were aged 3 to 4, and 41.4% aged 5 to 6 in Korea were using regular YouTube instead of YouTube Kids (Kim et al., 2020). The CAL programme developed in this study could also help children easily recognise what kinds of data might be collected and how they could be used to construct a YouTube recommendation algorithm and help children recognise rights over their data.

In our times of digital transformation and the discourse on intelligent information society, media education, at least in South Korea, seems to be still polarised between discourses of the dangers and opportunities that digital technology and media might cause and affect children's lives. In this context, media education seems to be regarded as a solution to prevent the dangers of children being exposed to cyberbullying and sexual and violent content while learning computational thinking, programming and artificial intelligence technology. However, in a datafied society in which our lives in public and commercial contexts are transformed into digital data stored, used and misused by technology giants such as Google, digital literacy should include critical understanding of data as part of data literacy (Polizzi, 2021).

The CAL program needs to expand media education to understand the digital media industry based on the concept of digital literacy from a socio-cultural perspective, including a critical understanding of the impact of digital media companies that process and use user data on individuals and society. This study showed that the pedagogy could be developed to raise awareness of critical understanding of YouTube algorithms for children and the need for critical algorithm literacy education as a new approach to media education in the age of the digital platform.

References

Andrejevic, M. (2007). *iSpy: Surveillance and Power in the Interactive Era*. Lawrence, KS: University Press of Kansas.

Auxier, B., Rainie, L., Anderson, M., Perrin, A., Kumar, M., & Turner, E. (2019). Americans and privacy: Concerned, confused, and feeling lack of control over their personal information. *Pew Research Center*. https://www.pewresearch.org/internet/2019/11/15/

americans-and-privacy-concerned-confused-and-feeling-lack-of-control-over-their-personal-information/

Bae, J.A., Cho, Y.H., Kim, S.S., Choi, E.H. & Hah, TH (2011). *A Study Measuring the Levels of Perception on Privacy (Protection of Personal Information) by Internet Users and Service Providers*. Korea Communications Commission.

Bae, S.Y., Lee, C.H., & Lee, J.R. (2020). *Youth media use and study on policy response plans by target I: Elementary student*. National Youth Policy Institute.

Bucher, T. (2017). The algorithmic imaginary: Exploring the expected effects of Facebook algorithms. *Information, Communication & Society*, *20*(1), 30–44. doi:10.1080/1369118X.2016.1154086

Buckingham, D. (2019). *The Media Education Manifesto*. Cambridge: Polity Press.

Couldry, N. (2020). Recovering critique in an age of datafication. *New Media & Society*, *22*(7), 1135-1151. doi: 10.1177/1461444820912536

Cooper, A. (1999). *The inmates are running the asylum*. Hoboken, NJ: Sams.

Choi, H., Jung, Y. & Nam J. Y. (2020). Users' Understanding of the Concept of Personal Information. *Korean Telecommunications Policy Review*, 27(2), 1–38.

Choi, M. (2016). A concept analysis of digital citizenship for democratic citizenship education in the Internet age. *Theory and Research in Social Education*, *44*(4), 565–607.

Cukier, K., & Mayer-Schoenberger, V. (2013). The rise of big data: How it is changing the way we think about the world. *Foreign Affairs*, *92*(3), 28–40. doi:10.1515/9781400865307-003

Gray, J, Gerlitz, C., & Bounegru, Liliana. (2018). Data infrastructure literacy. *Big Data & Society*, *5*(2), doi: 10.1177/2053951718786316.

Jacques, J., Grosman, J., Collard, A., Oh, Y., Kim, A., & Jeong, H. (2020). "In the Shoes of an Algorithm": A media education game to address issues related to recommendation algorithms. *The Journal of Education*, *3*(1), 37–62. doi: 10.25020/JoE.2020.3.1.37

Kim, Y.J., Kim, S.J., Lee, S.J. & Park, M.G. (2020). *Children and Media in Korea 2020*. Seoul: Korea Press Foundation.

Livinstone, S., Stoilova, M., & Nadagiri, R. (2019). *Talking to children about data and privacy online: research methodology*. London: London School of Economics and Political Science. https://www.lse.ac.uk/media-and-communications/assets/documents/research/projects/childrens-privacy-online/Talking-to-children-about-data-and-privacy-online-methodology-final.pdf

Livingstone, S. & Polizzi, G. (2019). *Data and privacy in the digital age: From evidence to policy*. Department of Media and Communication. *London School of Economics*. https://blogs.lse.ac.uk/parenting4digitalfuture/2019/12/11/data-and-privacy-in-the-digital-age-from-evidence-to-policy/

Napoli, P. & Caplan, R. (2017). Why media companies insist they are not media companies, why they are wrong, and why it matters. *First Monday*, *22*(5). doi:10.5210/fm.v22i5.7051

Pangrazio, L. & Sefton-Green, J. (2020). The social utility of 'data literacy'. *Learning, Media and Technology*, 45(2), 208–220, doi:10.1080/17439884.2020.1707223

Polizzi, G. (2021). Digital and data literacy: Comparing children's understanding of data and online privacy with experts and advocates data literacy practices. CILIP Information Literacy Group.http://www.hani.co.kr/arti/economy/it/909212.html#csidx738c4c9dfe410f3b65d946fdc9ccea9

Van Es, K, & Poell, T. (2020). Platform Imaginaries and Dutch Public Service Media. *Social media + Society*, *6*(2). doi:10.1177/2056305120933289

van Dijck, J. (2020). Seeing the forest for the trees: Visualizing platformization and its governance. *New Media & Society*. doi: 10.1177/1461444820940293

10

EMERGING FROM THE SHADOWS OF DATAFICATION

A favourable turn for cultural studies

Luo Xiaoming

Introduction

At the start of May 2020, Bilibili, an online streaming platform known for aggregating data on youth subcultures, launched a promotional short film called *The New Generation (《后浪》)*, specially produced for May the Fourth Youth Festival. In the video, a middle-aged man with the tone of an elder sings the high praises of the younger generation born in the new century. After being broadcast on multiple platforms, including the official CCTV, the film became a trending topic, and parodies subsequently appeared. In contrast to the highly praised younger generation, these short films treat the youth as "leeks" to be slaughtered, bearing the expense of the reforms led by their predecessors since 1990s. As such, some believe that the people behind this depiction of the new generation are the middle-aged adults, who reaped the benefits of economic reform. However, college students did not strongly oppose this interpretation; many still enjoyed and defended the film, despite knowing that it received lots of criticism. Others pointed out that WeChat reviews and classroom discussions are merely subjective impressions from a limited scope, representing a small group of people. Since then, most satirical short films have been blocked, trending topics have changed and discussions on the film have ceased.

This kind of phenomenon has become more and more common in China, a society that rapidly leapt into the world of the Internet. These hot topics do not lead to any questions or follow-up debates, but rather create more fog, making them all the more incomprehensible. This confusion of public opinion is not caused by datafication, but it is inseparable from people's full consciousness of the rapid expansion of datafication. One of the initial social functions of datafication is to show indirect relationships between facts in society. For example, the positive correlation between the suicide rate and the newly rich society was first discovered in *On Suicide* while using the statistical method in social investigation. When Durkheim used statistics from various European

DOI: 10.4324/9781003136842-10

countries to show that suicide was a societal problem, datafication laid the foundation for the establishment of "society" as a research object and created the ides of data contributing to positive societal attributes (Durkheim, 2006). Now, due to the high degree of datafication, it seems that anything can be recorded and stored. However, the social consequence is not making everything clear and transparent but actually producing more confusion and more uncertainty. For instance, in the example of the short film, there is data on the number of people who watched and shared *The New Generation*, the number of hours it was viewed, the viewers' occupations, age, gender, income, consumption level, location, social class and social interaction style. All these 'facts' exist as data and are automatically recorded somewhere. However, due to political control and business secrets, these data cannot be disclosed and shared with the public, let alone be included in analysis or critical thinking, which would give data its societal attributes and could promote the formation of a new holistic understanding. As a result, the massive amount of data that exists, but is always hidden from the public, creates what I call an increasingly looming 'shadow' over social consciousness leading to ordinary people's observations as separate from public debate.

This reality is embedded in the foundations of contemporary Chinese society and constitutes the starting point for this chapter's discussion on digital literacy and the role of cultural studies in addressing how society is created. Generally speaking, digital literacy is often divided into three different stages: acquiring "information age" skills, cultivating "habits of mind", and engaging in digital cultures and practices (Meyers et al., 2013). When discussing digital literacy in Chinese society under the dual control of a centralised government and profit-seeking capital, the first problem we face is how to define the engagement in digital cultures and practice, which form the basis for discussing which skills and habits of mind have yet to be cultivated.

Based on this principle, this chapter is divided into four parts. The first part outlines how, through the process of embedding itself into Chinese society, data and algorithms cause a set of shadows to pile up on the social mind. The second part explains the basic understanding and attitude of contemporary Chinese young people towards datafication, under this shadow. The correspondence and reflection between the two constitute the reality that cultural studies needs to address when thinking about digital literacy. The third part presents two cases that capture the thinking habits and blind spots of the younger generation in their understanding of datafication. Finally for the fourth part, given the above, the development of digital literacy in cultural studies is further implemented in the collection, sorting, filing, analysis, and comparison of "cultural facts". As such, the teaching methods, curriculum design and evaluation methods of cultural studies departments in mainland China also need to be developed.

What is in the shadow(s) of datafication?

With the increasing frequency of social movements on the Internet, the development of digital literacy is now seen as a third way to develop civic movements (Yue et al., 2019). In this process, Chinese youth shares many characteristics of the younger generation 'birthed by the Internet (网生一代)'. These characteristics include being increasingly distracted by the Internet (Hassan, 2012), the formation of database-like models of understanding (Manovich, 2002, pp.221–224), weakening independence due to increasing Internet usage (Lukianoff and Haidt, 2018, pp.181–190), the growing trend of consumerism, and constant discussions of politics in daily life (Arguelles, 2020), and so on. Unlike non-Chinese societies, Chinese youth developed these characteristics under a centralised Internet censorship system and big data monitoring, accompanied by the accumulation of distinctive shadows caused by them. In order to accurately grasp their characteristics and potential trends, it is necessary to further examine these shadows caused by datafication.

First, China's Internet censorship system creates a shadow of datafication on the social mind at a large scale. The Firewall, the review and supervision departments, the Internet administrators and whistle-blowers, those ordinary people with different opinions, positions and even tastes to show their disagreements, have step-by-step created a whole new censorship system, resulting in the social facts: texts, images, and videos disseminated on the Internet are "erased" on a daily basis. In this sense, China's Internet society has simultaneously undergone rapid expansion and reduction, constantly changing its landscape. Under a censorship system that is led and regulated by the government, and which has compliant civilian participants (regardless of their own wills), the criteria for deletion have become increasingly unclear through the combined force of political consciousness and algorithmic technology. Data that was once visible but became deleted now constitutes an increasingly vast, yet invisible existence. The more data is deleted, the less clear the criteria should follow, and the heavier the shadow in people's consciousness.

The second layer of the shadow caused by datafication originates from the process of collecting and using data by the officials at different levels. In Chinese society, the right to make and use data has always been in the hands of those with power, whether the decision-making will be data-driven or not depends on their wills. This process deconstructs the principle that every data point is equal since some data points are prioritised over others. This can be clearly seen in the collection and usage of personal information around the coronavirus pandemic in 2020. During the pandemic, all work units implemented a specially designed data system that required all personnel to report their personal information on a daily basis. They even collected teachers' and students' whereabouts directly through their phones' GPS systems. This type of data collection is compulsory, yet there was no guarantee or evidence that it helped protect against spreading the coronavirus or guarantee the individual right to free travel. For instance, during the first half of 2020, even in a metropolis like Shanghai, every resident could use the digital health code to take the subway, but they could not use it to freely enter neighbourhoods where they did not live. Similarly, after a local outbreak of coronavirus in Shanghai's Pudong District in October 2020, cities in other provinces refused to accept the effectiveness of Shanghai's health code and put forth requirements for quarantining and testing. The measures vary from city to city. The most severe of these requires all travellers from Shanghai to quarantine for two weeks or provide a five-day RNA certificate before entering the province. Even though the local government authorities had embraced datafication, they still disregard it in certain circumstances. How to formulate relevant data policies, all depends on what bureaucrats think is more conducive to controlling the areas under their jurisdiction and how easily they can shirk their responsibilities when problems arise.

This may be the primary principle for how datafication became embedded into Chinese society. In other words, the mandatory collection and superficial use of data is nothing more than a means for bureaucrats to achieve their goals and protect themselves in the current political climate. This means that in China, the datafication process under this type of high-degree control power has two parts: the first is the nearly borderless data collection, and the second is the process by which datafication is allowed to become embedded into social life. If we only recognise the former and overlook the latter, we cannot get a complete picture of the datafication process taking place in China. This two-layer datafication process is not only incapable of cutting down the random and arbitrary nature of power, but rather has become a new way for it to freely spread. Under this shadow, the social effects and potential problems caused by data and the algorithmic are almost secondary issues.

Finally, the unsurprising reality is that when the power to seize and use data is completely controlled by the government or large platforms, they are eager for data mining that would be helpful to consolidate power or to acquire more benefits. As Lev Manovich (2020, p.73) pointed out, commercial big data research often focuses on consumer behaviour data and ignores others. This is particularly evident in China at the moment. This ensures that parts of real life that have nothing to do with this are either not recorded or excluded due to conflict with those in power's goals and interests. Since the second decade of the new millennium, Chinese society is rapidly urbanising and becoming highly digitalised, yet great tracts of reality are not recorded or are deliberately erased. One is to make it easier for people to live in the "eternal present" (Hartog, 2015), which means the present cannot open to the past out of control or the future connected with uncertainty, according to the will of power through the support of algorithms. The other is the third level of shadow located in social consciousness with fewer feelings and reactions about the realities deleted by datafication, since more and more young people are accustomed to living in the digital world. It can be thus said that the shadows produced by datafication and algorithmic technologies in the new Chinese society are constituted at three levels. The way people understand and grasp the reality of datafication stems from an accumulation of shadows.

The structure of feeling in the shadows of datafication

Chinese youth, who are also under these shadows, have formed a corresponding structure of feeling that they use to understand data, digital technology and their relationship with datafication.

First of all, contemporary youth who make up the main force of the Chinese Internet grew up immersed in Baidu, Taobao, Weibo and WeChat. According to the

45th CNNIC Chinese Internet Development Statistical Report (2020), as of March 2020, there were 904 million Chinese Internet users; among them, Internet users aged 20–29 made up 21.5%, and Internet users aged 30–39 made up 20.8%, greater than other age groups. In theory, the young generation should be the group that is most aware of datafication. However, when technology has always been distorted by power, it is easier to give birth to two modes of understanding datafication, created out of disgust and avoidance. The first mode, in opposition to an arbitrary power, is that they speculate datafication and technology exemplify objectivity and fairness. The second mode is an inability to distinguish the boundaries between power and technology, preventing them from deeply studying the relationship between the two. Regardless of the mode, arbitrary power will always be an obstacle to serious critiques of digital technology. Social monitoring and other problems brought about by datafication, including the system of daily algorithm recommendations (Zuboff, 2019), are hidden from vision.

Secondly, contemporary youth grew up with Internet censorship, exemplified by the Great Firewall, which is the combination of legislative actions and technologies enforced by the government of China to regulate the Internet domestically. To them, this wall, which has existed for as long as they have been alive, is as natural as air; only under special circumstances or opportunities do they become aware of its existence. (Denyer, 2016; Li, 2018). This is an important condition for the formation of Chinese society's method of producing and understanding data. At the same time, due to both the "long history" of the Great Firewall, and the so-called "Rise of China", the younger generation's views on the Great Firewall are not always negative. They usually think China is a developing country and needs social stability. It calls the necessity for the Communist Party to make a line or build a wall to make sure that the state media and school textbooks could repeat their will all the time (Li, 2018). This is the backdrop against which Chinese people are acquiescing to increasingly strict Internet censorship. The ways in which the Firewall represents the arbitrariness of the political censorship system, and the blurred but rigid awareness of self-censorship, have become the norm. There is no clear standard for what types of websites should be blocked and what kinds of information should be filtered. The power of interpretation is left entirely in the hands of the government and cannot be discussed by the public (Reed, 1999). It is under these circumstances that the younger generation who grew up with the Great Firewall have essentially cultivated cultural inertia; they do not have the passion to delve into standards or principles and believe that they are unable to find them at all.

At the same time, the Great Firewall is not absolute. The younger generation has always been better at adapting to technology. If they want to circumvent the Firewall, the world no longer seems to be out of their control. In order to deal with Internet censorship, Chinese youth have developed a common habit of saving and downloading things left and right. This mode of participating on the Internet creates two illusions. The first illusion is that they are facing a perpetually cracked wall that does not act effectively as a barrier. In an 18-month survey conducted by Chen and Yang of 1,800 Beijing college students, the first conclusion is access to uncensored Internet alone has little impact on students' acquisition of politically sensitive information. Nearly half of the students do not use the tools to bypass censorship at all (Chen and Yang, 2018). And students brought up another aspect of this illusion often during class discussions – the Firewall cannot stop them and does not make them feel like they lack freedom of information, so it is not a big problem. The second illusion is that saving, downloading and sharing is a form of resistance by free Internet technologies. Due to the proliferation of information on the Internet and the ability to circumvent the Firewall, these two illusions have significantly released the pressure created by datafication.

Finally, the individual and data-based facts formed on this basis, as well as the relationship formed between the individual and society, have established a kind of peace and security amidst extreme alienation. The selectivity of the Internet (Manovich, 2002, pp.123–128) has evidently played a role in maintaining this type of attitude. On the one hand, in our micro-world, the youth seem to have endless choices – this is exactly what the short film, *The New Generation* celebrates. But on the other hand, once they face "reality", they often believe that "reality" is difficult to change, and that their own choices are meaningless and cannot affect society.

For contemporary Chinese youth, their inability to think deeply about technology and data, the normalisation of the defects of competitive truths, their lack of clarity on the object of resistance, their fragmented sense of history and reality, and the extreme degree of alienation between individuals and society all constitute important features of the structure of feeling in datafication. Mainland Chinese cultural studies must confront this structure of feeling (Williams, 2011, p.69) in analysing datafication.

Narrative loss in negative liberty: two cases from a masters' programme

The spring term in 2020 was the first term that all University courses had to be organised online during the coronavirus pandemic in mainland China. At that time, no one could know when lockdown would end, and everything seemed to happen online. This turn in course delivery took aim at the ability of cultural analysis to offer insight into online social issues at that time. This course was a professional course for the first- and second-year master's degree students of the Department of Cultural Studies at the University of Shanghai. About ten people participated in the course for each iteration. From this Masters' programme, I offer two prototypical cases as representations of the structures of feeling of the younger generation when it comes to the Internet.

The first case focuses on young people's ability to respond digitally to the coronavirus pandemic in the face of Internet censorship. Starting in late January 2020, with the outbreak of the coronavirus in Wuhan and the ensuing lockdown of the city, Internet censorship in China reached a peak. WeChat's public accounts are a good example. A huge quantity of content either failed to pass background review or was immediately deleted after being posted. The most well-known example is the article on the Wuhan coronavirus "Whistle-blower" – Ai Fen (艾芬). This article was quickly deleted after being published on the "People" public account, and the account was simultaneously frozen. Filled with rage at this official blockage of the news, people began implementing various strategies to rewrite or translate the deleted article, such as using dialects, pictographs, IOB (inscriptions on ancient bronze objects), li calligraphy, handwriting and morse code to evade online censorship. Different versions of this article provided a modern spectacle of spontaneous relaying and information dissemination (Sheng, 2020). However, using such prominent and eye-catching methods to record and distribute one article to keep it online meant that only a minority of articles deleted by the censorship could be recorded in these ways. Larger amounts of information and comments were being deleted within a few minutes or days of being posted, which happened daily during the coronavirus pandemic. Therefore, an experiment in the Masters' class invited students to record a month's worth of deleted WeChat public account articles. Through this experiment, young people were forced to react to "deletions" and started to observe their own actions.

After a month, the records that students submitted shared several common characteristics. First, the article titles and author's names were all preserved. But most fields regarding which public accounts published the articles, the dates they were published and when they were deleted, the reasons for being deleted, let alone tracking down the deleted articles, were empty. Nevertheless, one article deleted from a public account may still be posted on another public account in a different way. Tracking can help figure out what level the order to delete came from, whether it be a spontaneous whistle-blower, a self-censoring platform or an order from an authority. Next, most students only passively recorded deleted contents on official accounts they already followed in their WeChat or Moments, and displayed no awareness of the gap between the censorship driven by keywords searching and the deletions made by humans. There was little recognition by the students of their incomplete, filtered perspectives. To them, their records were complete. Finally, since they did not recognise that these records were not necessarily complete, the students did not consider any ways to supplement their work. For instance, few students thought to track and record a specific topic, which would be a chance to note the difference between data-driven or human decisions in this censorship system.

During the class discussion about this experiment, the students were surprised that they had completely ignored these "details". Yet this negligence demonstrates their basic understanding of the relationship between deleted digital content and their files of digital records. In other words, the students were occupied with recording and preserving, but spent no time thinking about how this digital content should be sorted, marked, archived and supplemented, let alone actively reflect on what types of digital records they should strive to create and what kind of censorship system they need to face as the whole. Unfortunately, this is a common reaction for young people when they face the Internet censorship system.

The second prototypical case is "Writing Generator", a mini programme proposed as an object of study by the students under the theme "The plights of Language in the Internet Age". During the pandemic, major colleges and universities held online pandemic prevention classes and required students to write about what they learned after each lecture. Most students were reluctant to write such essays, in which they are ordered to record correct views and show positive emotions, as this did not connect with their thinking and living. As a result, someone launched a free programme called "Pandemic Prevention Class Impressions Generator" to help students automatically generate content. According to reports, about 350,000 students used this programme to generate afterthoughts about the pandemic prevention class (Chen, 2020). At the same time, there was a special "writing generator" nicknamed "Bullshit (胡说)" casting innuendoes against Hu Xijin (胡锡进), who was the chief editor of *Global Times* (环球时报) as his comments always incited resentment among the younger generation during the pandemic. For them, Hu Xijin always says the right words and makes no mistake. Yet, during the COVID-19 time, nationalism is emerging among the 'netizens'. The position of Mr Hu's comments is too subtle to take. Neither the nationalists nor the anti-nationalists would like his articles and comments. To use this writing generator, one only needs to enter keywords like "nation", "United States", "Hong Kong" and so on, and the program- automatically generated Hu Xijin-style political commentary or editorials.

Not surprisingly, the pandemic prevention class afterthoughts written by the mini-program seemed full of positive energy, drowning in the ocean of jargon. The comments of Hu Xijin appeared to express political opinions, but all were empty rhetoric. The students' assessments of them were all the same. These two comments were typical:

> One peculiarity of "Hu Xijin-style" is the subjects of these sentences are usually unclear. The meaning of it is ambiguous... The sentences always lack the actual subject. The readers only get some macro, abstract and irresistible term such as people, China, State and so on from his comments.
>
> (Wu, 2020)

> I think [the generator] is funny and I have played them a few times. There are many similar automatic generators, such as review article generators, thought report generators, nonsense article generators and even essay generators. The article written by these writing programmes may look great but you don't know what they're talking about... As for the content that it generates quickly, it is naturally formalistic.
>
> (Liang, 2020)

In general, they regarded such programs as a convenience for everyone, a joke, and enjoyed the implicit political stance full of contempt and mockery of official ideology. It could be said that this is the Internet generation's unique way of responding to the empty formalism of 'Chinese characteristics'.

In these two cases, the basic attitudes of the younger generation towards datafication are quite consistent. Just as the comment made by one of the students they are driven to believe in cynicism:

> But the interesting thing is that these clichés are not meaningless. If you generate a few articles and observe them, you can get many familiar logics which are very used in the current media propaganda and even public comment… This may reveal the state of "believing in cynicism", but it is also rationalizing this state of "believing cynically".
>
> (Liang, 2020)

In other words, they actively use what Internet and technological knowledge they have to maintain their limited and shrinking negative liberty. In this regard, the negative liberty that Berlin (2002) writes about – freedom from doing something – has been further rewritten as the active use of algorithms and data to protect freedom from doing something. In the first case, preserving screenshots and recording the names of the titles and authors of deleted articles has become an instinctive reaction to Internet censorship. This response, however, is a passive preservation of digital content at risk of being snatched away, and does not consider the key question of what type of approach is truly effective in the face of systematic information deprivation. However, they did not ask what, beyond fragmented information and contents, is the real thing they need to protect and cherish by preserving and recording. In the second case, mocking the formalistic writing procedure fully exposed the glaring reality of the high degree of disconnect between language and reality. However, while saving students from the pain of writing meaninglessly, this procedure also rendered the key question – the relationship between language and reality, between the writing and own feelings and thoughts – irrelevant through the method of automation. The freedom of private possession after saving screenshots and the freedom of satire brought about by the automatic writing programme cynically are still both forms of negative liberty.

In addition, these two cases also highlight an even more common problem with the datafication process. As database logic increasingly occupies a dominant position, narrative logic is completely lost. In the first case, when faced with missing data, the students adopted database logic without thinking. However, "a database can support narrative, but there is nothing in the logic of the medium itself which would foster its generation" (Manovich, 2002, p.201). Data that appears to be actively preserved does not actively present narrative logic and organisational principles. The second case goes a step further. Through algorithms, database logic disguised as narrative logic began to appear. Everyone could use the Writing Generator to appear to demonstrate unique written afterthought. Trying to describe their own experience during the pandemic time seems neither necessary nor convenient and might also pose a risk. To accept the narrative logic disguised by the disguised database logic is a more popular choice in the younger generation.

As Manovich (2002, p.199) pointed out, "Databases and narratives are natural enemies. They are competing over the same territory of human culture. Each claims to have an exclusive right to make meaning out of the world." If the biggest change brought about by the Internet as a form of media is that database logic is increasingly dominant, then a closely related issue in contemporary culture is: Will those clichés, which were originally written by persons and now by the generator online automatically, be reproduced more and ore in the digital society with the support of algorithms? Will the shadows of datafication accumulate further through automatic writing programmes? Will they replace the thinking of humans? At that time, what does narrative logic mean? Should it not be usurped or arbitrarily imitated by database logic if we consider its significant political and cultural significance in the shadow of datafication?

It seems that the use of positive digital techniques to embrace negative liberty and database logic is the instinctive reaction of the young generation. What teachers of cultural studies need to imagine is the next step they should take to replace these instinctive reactions to the shadow of datafication. When reconsidering the reactions described above in the Masters' course, it is not hard to discover that the teachers asked for more explicit requirements on digital literacy. It also constitutes the starting point for cultural studies to focus more on "cultural facts".

Establishing "cultural facts"

The uniqueness of the shadows cast by the digital, their effects on misapprehending the social world, how they create structures of feeling and the resulting reactions by the younger generation argue for a need for digital literacy. The goal of digital literacy in the Chinese context involves breaking the vicious cycle between datafication and the arbitrary use of power, disrupting the instinctive reaction towards the shadow. During this vicious cycle, being short of the basic facts, the young generations are losing their capacity to make judgements, understandings and the ability to think critically about datafication. The power to decide what kind of social facts get turned into data is highly centralised. It is more and more unavoidable, trapping people – especially youth – in a state of powerless cynicism where the state controls everything.

Therefore, digital literacy in mainland China should include certain elements. Firstly, helping the young generation recognise the digital world they live in, which consists of not only the automatic data collecting process which reduces everything into one speciously data-driven decision-making society, but also the digital shadow that comes with this process and the reactions against it. Secondly, recognising meaningful social phenomena and translating them into forms of datafied expression, are all significant abilities when living in the digital world. An ability to read data inscriptions closely helps people to understand how they are created and represent significant social phenomena. In other words, digital literacy will help to clear the shadow that darkens social consciousness. If every student is aware of the collection and use of their travel and health data by 'the online system of their university, they would have a clear thought that this process has the more symbolic meaning of surveillance than a health control effect. Then they would find most steps of those official data gathering in this society follow the same social logic. The myth of data claimed by the official and the promise they made about it would be easier to break. They need to gain and distinguish the social meaning of data by themselves. Finally, there are concrete challenges. How to redefine archiving during the process? How to confirm the boundary between the narrative and database logic to keep balance in the digital world? How to scan the blind points hidden in the reflex reactions? And so on. But above all is the ability to put data-driven process into cultural evaluation. In mainland China the discipline of cultural studies helps us focus on the significance of 'cultural facts' to advance digital literacy.

Understanding social facts as "objects" is the foundational rule by which Durkheim (1982) established sociology. Lev Manovich highlighted the new problems facing cultural analysis in the new century:

> In other words, what is unique in cultural analytics is not what you do with data – *but how do you get from this elusive thing we call culture to its data representation in the first place?* [emphasis added] That is, how do you turn cultural experiences, events, actions and media into data? What is gained and what is lost in this translation?
>
> (Manovich, 2020, p.73)

In his view, cultural analysis needs to recognise that objects today are shaped by the digital (p.10). Data does not produce the new representation directly but has been the basic unit to control the process of representation. The distinction between Durkheim and Manovich is located in their attitudes towards sources of power to define and regular representation. For Durkheim, the decisive power came from the social mind, which belonged to someone or some groups. For Manovich, the algorithms crucially decide media visualisation. There are algorithms beyond the individuals to reveal or grasp those facts and discover cultural models they overlook unconsciously. Manovich noticed the warning of "data variously 'cooked' within the varied circumstances of their collection, storage, and transmission" (2020, pp.385–386), and also confirmed the importance of questioning existing metadata and indeed destabilising the existing categories encoded in institutional metadata. Yet, the task of defining and encoding the realities through digital technology set by him seems too optimistic to delete or shrink the shadow of datafication in Chinese society. The case of the "Writing Generator" reveals this challenge. If researchers in the future try to study what kind of feeling ordinary people had during the pandemic, how can they make a distinction between the afterthoughts written by humans and by the writing generator? Furthermore, how can they finally get the idea that neither afterthoughts written at the time, nor the contents online are the right clues for them to discover the experience of this pandemic? In this case, the plight is not that the data would disappear over time, but that the social mind could not filter, re-examine data and evaluate social phenomena.

The right to make judgements has been ceded to algorithms but is still controlled by power. In other words, how to make judgements and to do so by collecting data according to its social meaning is the challenge we need to face directly. Here, to archive realities into digital facts and to analyse them through those designed algorithms has to be the first step. Yet, it is not enough for the persons and the educators who care about cultural progress. "Social stability" is a noticeable Chinese phenomenon in the academic circle, which often regarded as a self-evident national goal in social science in China (Chen, 2019). Social facts need to be recognised as cultural facts in this digital era to keep a necessary distance from the discourse of social stability. This will help us to locate our place and evaluate the meaning of data or algorithms within the development of humanity as a whole (Mills, 2000, p.6). During the expansion of datafication, everything will be datafied sooner or later, and this creates a new culture – as defined by Raymond Williams (2017, p.387):

> The idea of culture is a general reaction to a general and major change in the conditions of our common life. Its basic element is its effort at total qualitative assessment... General change, when it has worked itself clear, drives us back on our general designs, which we have to learn to look at again, and as a whole. *The working-out of the ideal of culture is a slow reach again for control.*
>
> [emphasis added]

In this respect, statistical literacy, data visualisation, and the ability to question data, are the necessary foundation for cultural analysis, but far from enough. The prime task in contemporary Chinese society is to achieve a social definition of data and algorithms by gathering cultural facts to release them from the absolute domination of politics and capital, where digitising a certain part of social reality, naming it, giving it social attributes by recoding, has become the norm.

For the young generations, what they have been deprived of for a long time is not access to new digital skills, but the way to think of them and what they can do within the world. In this respect, to collect, achieve and analyse "cultural facts", not only aims to explore the above-mentioned social conditions that fully naturalises the digital, but also encourages young people to shift themselves away from passive self-protection. It challenges young people's awareness of data and structures of feeling in two ways.

First, cultural facts require students to (re)capture, identify, and classify digital things in relation to their questions which are rooted in their social experiences and confusion. This part of the work can take advantage of existing digital facts, including images, text and videos online, but it is impossible to rely on them. Any transformations concerned with cultural reconstruction will usually be affected or distorted by the shadows of datafication during their investigation, research and thinking. As such, the classification and labelling of digital facts by various platforms and the resulting digital objects must undergo a process of decoding and re-archiving under the eye of both technology and cultural values. During this process, digital technology is bound to depart from the route predetermined by political power and capital interests and be used relatively independently.

Secondly, to a certain extent, an ability to collect and interpret data according to a narrative logic instead of the database logic can protect the individuals against the data-driven decision-making world dominated by political and economic power. There are two reasons for this ability. One is mainstream narrative logic occupied by state power or capital usually hides and disguises as database logic in our Internet Age. The other is the individuals and social groups' ability to collect data in China is further limited by the government or tech company, placing individuals in a powerless position. Interpreting data through a different narrative logic will help individuals to improve their power relative to such entities. Digital literacy should therefore include the ability to collect and make data according to narrative logic in this context.

Finally, if cultural facts have been built using this approach, young people will also find a middle ground between active technological use and passive self-protection. This middle ground will assist them to reflect and refresh their understandings of resistance. Social intervention in the digital world will also begin to emerge during reflection. Because the research work of sorting, preserving and accumulating data into cultural facts, is not only to broaden the perspectives of cultural studies, but also to avoid popular cynicism. Young people can improve their critical thinking abilities and form their interests by preserving different social memories and realities. If this argument has some validity, then cultural studies need to change existing teaching and evaluation methods.

First of all, from curriculum design to specific teaching methods, the pedagogy of cultural studies should focus on strengthening the abilities of digital literacy, competing for and accumulating "cultural facts", and building a supportive academic platform. In this third decade of the new millennium, such a change of pedagogy could be a new way for cultural studies to take its anti-system stance within the Chinese university system. Secondly, in terms of evaluation methods, current cultural studies still place too much emphasis on opinion or original ideas, while demonstrating insufficient understanding and underestimation of the significance of digital technology, classification, preservation, and accumulation of "cultural facts". As a result, theories, particularly Western theories, are still hard currency in academic exchanges. Students would rather hide in the library than conduct actual investigations and research and do the work of identifying and accumulating cultural facts in the world. In response to this situation, we need to revise the current academic evaluation method, which places too much emphasis on insights, and establish a new standard of using "cultural facts" as the new currency for cultural research and the humanities.

For example, the health code system promoted by the Chinese government during the COVID-19 pandemic has become the research object in different disciplines. The unique research supplied by cultural studies should be the cultural and social effects on ordinary people using this kind of health code in everyday life. How to collect the experiences of using the health code? It might include those words people launch on Weibo or WeChat to express their feelings. However, the actual ways of use between different cities, regions, shopping malls, public areas and even universities, all the concurrent controls in various units are important parts of feeling and experiences that students need to follow and collect. Since the social and cultural effects produced by these overlapping digital systems is a significant means to understand the digital mode of governance after the pandemic in China, the way to trigger people's psychological and cultural responses need to be sampled in this kind of collection. Then further analysis could happen on this basis. In the same way, to discuss the plight of language in the Internet Age and take Writing Generator as a case, just by commenting or sketching the cynicism in it is far away from enough. The facts that need to be clarified should include: how many similar writing generators launched, what kind of purposes are they used for, what consequent impressions do people have after using them, and will there be any fields they refuse to use automatic writing programmes forever, etc? You could get these pieces online, on news reports or by face-to-face interviews, but none would be recorded consciously by the datafication. All these constitute the object that cultural studies need to investigate, record, categorise and follow up and review when drawing the current social and cultural map.

Finally, this change should become a requirement for students' dissertations. In addition to submitting the discussion portion, they should also include a database of "cultural facts" to support their discussion. For the first part of the dissertation, academic journals might publish them, or the online library will collect them. For the database of cultural facts, students should save and share it in a public or semi-public online platform, so that these facts can be stored, used, tested, compared and updated by subsequent researchers.

The datafication of the world appears to have distorted and constrained social discussions, which has been aided by the young generation's increasingly shorter memories and seemingly unlimited interference (Hassan, 2012). When the timeline of social life and individual experiences are constantly folded on top of each other, cultural creation, accumulation and research faces unprecedented challenges. Under the shadows of datafication, this crisis in contemporary Chinese society is particularly striking. The more we recognise power's ultimate ability to change the definition of culture, the more we fully understand the decisive force that governs the current datafication process. It is not difficult to draw this conclusion: what truly needs earnest reflection and careful financing is how to face the shadows, search and integrate new motivations in the structure of feeling of contemporary young people, and allow cultural studies and allied disciplines to make full efforts to capture, define, and organise "cultural facts". This would be a better approach to raising and redefining digital literacy in the context of China and formulate more practical tactics.

References

Arguelles, C. (2020) 'Is youth really wasted on the young? Digital citizenship as an emerging politics of Filipino millennials,' *Refeng Xueshu*, 10 April [online]. Available at: http://www.cul-studies.com/Article/view/article_id/369.html (Accessed: 2 Dec 2020)

Berlin, I. (2002) *Liberty: Incorporating Four Essays on Liberty*, Oxford: Oxford University Press.

Chen, Y. F. (2019) 'What is the "Social Stability"? – Discussion of Concept and Research Methods,' *Tsinghua Journal of Social Sciences*, 1(1), pp 66–85.

Chen, Y. Y. & Yang David Y. (2018), 'The Impact of Media Censorship: Evidence from a Field Experiment in China,' [online]. Available at: https://stanford.edu/~dyang1/pdfs/1984bravenewworld_draft.pdf (Accessed: 11 Nov 2020)

Chen, Z. H. F. (2020), '350,000 people use the writing generator to automatically write the afterthoughts of the pandemic prevention class Netizen said it was the same level if I did it myself,' *Nanfang Metropolis Daily*, 12 March [online].Available at: https://www.sohu.com/a/379604606_161795 (Accessed: 12 Dec 2020)

Denyer, S. (2016) 'China's scary lesson to the world: Censoring the Internet works', *Washington Post*, 23 May [online].Available at: https://www.washingtonpost.com/world/asia_pacific/chinas-scary-lesson-to-the-world-censoring-the-internet-works/2016/05/23/413afe78-fff3-11e5-8bb1-f124a43f84dc_story.html (Accessed: 10 Dec 2020)

Durkheim Émile.(1982) *Rules of Sociological Method*, New York: Free Press.

Durkheim Émile.(2006) *On Suicide*, London: Penguin Classic.

Hartog, F.(2015), *Regimes of Historicity: Presentism and Experiences ofTime*, New York: Columbia University Press.

Hassan, R.(2012) *The Age of Distraction: Reading Writing, and Politics in a High-speed Networked Economy*, New York: Routledge.

Li, Y.(2018) 'Young People in China Don't Know the Internet We Do – And They Like it That Way,' *Independent*, 05 Sept [online]. Available at:https://www.independent.co.uk/life-style/gadgets-and-tech/features/china-internet-social-media-great-firewall-china-censorship-apps-a8510036.html (Accessed: 10 Dec 2020)

Liang, C. L. (2020). *Sharing Note on the course*, 21 Apr.

Lukianoff, G. and Haidt, J.(2018), *The Coddling of the American Mind: How Good Intentions and Bad Ideas Are Setting Up a Generation for Failure*, London: Penguin Press.

Manovich, L. (2002), *The Language of New Media*, Cambridge: MIT Press.

Manovich, L.(2020) *Cultural Analytics*, Cambridge: MIT Press, Apple Books.

Meyers, E. M., Erickson, I. and Small R.V.(2013) 'Digital literacy and informal learning environments: An introduction,' *Learning, Media and Technology*, 38(4), pp 355–367.

Mills, C.W.(2000), *The Sociological Imagination*, Oxford: Oxford University Press.

Reed, K. M.(1999), 'From the Great Firewall of China to the Berlin Firewall: The Cost of Content Regulation on Internet Commerce,' *Transnat'l Law*, 12, 543 [online]. Available at: https://scholarlycommons.pacific.edu/globe/vol12/iss2/9 (Accessed: 15 Nov 2020)

Sheng, Y.(2020) 'Pneumonia Epidemic: "Whistle-blower" Triggers an Anti-Censorship War, the Chinese Strike Back Creatively', *BBC*, 11 March [online]. Available at: https://www.bbc.com/zhongwen/simp/chinese-news-51831652 (Accessed: 12 Dec 2020).

The 45th CNNIC Chinese Internet Development Statistical Report.Available at: http://www.199it.com/archives/1041487.html. (Accessed: 21 Dec 2020)

Williams, R. (2011), *The Long Revolution*, Cardigan: Parthian.

Williams, R. (2017), *Culture and Society*, London:Vintage.

Wu, Z.W. (2020). *Sharing Note on the course*, 21 April.

Yue, A., Nekmat, E. and Beta, A.(2019) 'Digital Literacy Through Digital Citizenship: Online Civic Participation and Public Opinion Evaluation ofYouth Minorities in Southeast Asia', *Media and Communication*, 7(2), pp 100–114.

Zuboff, S.(2019) *The Age of Surveillance Capitalism: The Fight for the Future at the New Frontier of Power*, New York: Public Affairs.

11

CHILDREN'S PRIVACY AND DIGITAL LITERACY ACROSS CULTURES

Implications for education and regulation

Sonia Livingstone, Monica Bulger, Patrick Burton, Emma Day, Eva Lievens, Ingrida Milkaite, Tom De Leyn, Marijn Martens, Ricarose Roque, Katharine Sarikakis, Mariya Stoilova and Ralf De Wolf

Introduction

Digital technologies increasingly provide the infrastructure for education, political and cultural participation, work and family life. This makes them a crucial means by which people can exercise their rights and have their needs met. One such right, to privacy, is particularly threatened by the adoption of new forms of data collection, processing and surveillance enacted by businesses, the state (including education, health, law enforcement and welfare systems) and the general public (including the family). Recognising children's rights as set out in the UN Convention on the Rights of the Child (UN (United Nations), 1989), and their relevance to digital technologies as explained in the UN's *General Comment No. 25 on Children's Rights in Relation to the Digital Environment* (UN Committee on the Rights of the Child, 2021), this chapter examines the implications of digital technologies for children's privacy, both as a human right and as a means of exercising other rights in a digital world.

A child-rights approach focuses on the responsibilities of a government as the primary duty bearer. While there are many actions open to governments, protecting children's privacy has become, in part, a task for educators, newly charged with explaining the digital world to their students so that they can, supposedly, protect themselves. From the basics of access to and operation of devices through to a critical grasp of how personal data is processed, digital literacy is crucial to inclusion, equality and other rights in a digital age. For this reason, and to promote the full range of human rights in relation to the digital environment, digital literacy is a policy priority in the EU (Carretero et al., 2017) and internationally (Nascimbeni and Vosloo, 2019), recognised as important for a host of democratic and social justice outcomes. However, the imbalance in power between platforms and individual users sheds a critical light on the expectation that the right to privacy depends on individuals

DOI: 10.4324/9781003136842-11

taking on the responsibility for acquiring digital literacy. Such a burden is even less appropriate when it comes to children. Consequently, protecting children's privacy is also a task for legislators and regulatory authorities, charged with establishing and enforcing privacy-respecting frameworks within which the government itself, as well as businesses and other actors, including individuals, should act.

Neither the educators' nor the legislators' task is easy, and in many parts of the world, each has hardly begun. Moreover, in ways that are rarely examined, they are in certain ways interdependent. On the one hand, privacy and data protection regulations work more effectively when data subjects (that is, users, children, the public) are educated to be informed, critically aware and capable of giving meaningful consent. On the other hand, education is best facilitated when its subject matter is knowable – orderly, transparent, documented and regulated. At present, most people, children or adults, have a poor understanding of how digital systems work, including the commercial data ecology. This poor understanding is both cause and effect of people's relative exclusion from opportunities to transform, generate or intervene in such systems (van Dijk and Hacker, 2003; Chakravartty and Sarikakis, 2006). The problem is exacerbated by the fact that regulatory solutions (such as requirements for transparency and accountability of platforms and data brokers) do not work well in practice and are insufficiently enforced (Milkaite and Lievens, 2020). During the COVID-19 pandemic, the lack of public understanding regarding data processing played a key part in the confused debates regarding the surveillance potential of government-sponsored health-tracking, the legitimacy of the data-mining strategies of the educational technologies rolled out for home learning, or the difficulties in preventing 'anti-vax' and other disinformation.

In this chapter we examine the interdependences between educational and regulatory solutions for children's privacy in a digital world. We ground our analysis in our qualitative research with children in diverse countries exploring how they perceive privacy as an idea, how this in turn manifests in their privacy practices, and the changes they call for, before suggesting ways in which education and regulation could be more responsive to children's needs.

The co-evolution of education and regulation approaches

Both educational and regulatory 'solutions' to digitally mediated threats to children's privacy have their particular histories and legacy. Digital literacy initiatives, the 'heir' of the technological literacy debates of the 1990s (see, for example, Fulton, 1997), as well as the media literacy debates of preceding decades (Buckingham, 2007), have co-evolved with the technological innovation. Digital literacy initially focused on the access and resources that inhibited or enabled people to be technologically fluent (and affluent). Although research has since expanded expectations beyond the functional use of technological interfaces, the critical rethinking needed to overcome instrumentalist approaches and encompass the dynamic spectrum of critical capabilities required for digital and social justice (Aviram, et al., 2006; Raffaghelli, 2020) has only partially been realised (Livingstone et al., 2008). While in the past it could be expected that people could gain a degree of critical knowledge of media and information systems, the complexity of today's technological context is increasingly challenging.

In short, early digital literacy education developed for a simpler technological world and has not kept pace with innovation or regulation. The level of literacy demanded in today's technological complexity, even to manage the interface (for example, by adjusting or changing default privacy settings) requires a significant investment of time, let alone to grasp the privacy implications of technological design or the data ecology. Moreover, the economic and educational environments in which children are growing up are highly variable cross-nationally, resulting in considerable inequalities in the educational resources available to children for digital access and digital literacy (Nascimbeni and Vosloo, 2019). Expecting a thorough understanding of digital environments from ordinary citizens, especially children, would be a serious error in regulatory thinking.

Since 2000, the most influential legislation internationally has been the US Children's Online Privacy Protection Act (COPPA), which established the data collection regime under which the major US companies (notably, Facebook, Apple, Amazon, Microsoft and Google) operate regarding children. COPPA imposes certain requirements regarding notices and collecting verifiable parental consent on operators of websites or online services that direct their services to children under the age of 13 or who have actual knowledge that they are collecting personal information from a child under the age of 13. The European Union's (EU) regulatory response has been far wider reaching in scope, applying to all EU citizens. The individual's right to data protection being recognised as a separate fundamental right (Article 8 of the EU's Charter of Fundamental Rights, 2012), the EU established specific legal requirements and obligations to realise this right in the General Data Protection Regulation (European Union (Parliament and Council), 2016). This is possibly the strictest data protection framework internationally and is influential well beyond Europe. The GDPR makes provision for children in Recital 38:

> Children merit specific protection with regard to their personal data, as they may be less aware of the risks, consequences and safeguards concerned and their rights in relation to the processing of personal data. Such specific protection should, in particular, apply to the use of personal data of children for the purposes of marketing or creating personality or user profiles and the collection of personal data with regard to children when using services offered directly to a child.

Significantly, the rationale for extra protections for children is their presumed awareness regarding uses and abuses of their personal data. This represents but one of many challenges facing both those who process children's data and those who need to enforce the GDPR. Beyond Recital 38, references to children throughout the GDPR are few, and provisions are notoriously vague. Looking more widely, legal contexts around the world leave considerable room for interpretation for actors processing data in regard to their specific obligations towards child data subjects (Milkaite and Lievens, 2019). In this context it is important to note that the USA jurisdiction refers primarily to 'data privacy' while the European regulatory frameworks set up 'data protection' rules. In this chapter, we refer to privacy as encompassing both data privacy and data protection, since 'privacy' is the notion that children use most often, including when they discuss the collection, processing and control of their personal data in digital contexts. Where we explicitly discuss the regulation thereof, we refer to data protection.

Including children in matters that affect their lives

Today's children are the first generation to grow up with a ubiquitous and omnipresent reliance on digital technologies for learning, communication, health, leisure, work and politics. Hence the imperative for society to attend to the rights and needs of its young citizens in relation to the digital environment, as set out by the UN Committee on the Rights of the Child (2021). Article 12 of the UNCRC concerns children's right to be heard on all matters that affect their lives, and to have their views given due weight by relevant decision makers. Yet children are rarely invited to participate in formal consultations on the development of legislation and policy relating to education or regulation, and they have little opportunity to co-construct the processes and environments that affect them, although child rights advocates have developed effective methodologies for such purposes (Lansdown, 2014).

Only recently has literature sprung up asking what children understand about their privacy in digital contexts and what changes are needed to realise their rights (Stoilova et al., 2019; De Wolf and Vanden Abeele, 2020). Children's lives differ on many dimensions, and particularly relevant to their privacy are familial and cultural values, education (especially digital literacy curricula), public debate about technology and trust in government. Using a common methodology that combined research and consultation with children, this chapter explores children's voices and perspectives on their privacy and the uses of their personal data in relation to the digital environment in very different parts of the world – Austria, Belgium (Flanders), Cambodia, Indonesia, Malaysia, Thailand, the UK and the USA. In each context, the researchers began by asking children about their online activities and their unprompted perceptions of privacy and digital practices (such as selecting apps, checking age restrictions, reading terms and conditions and changing privacy settings). We then discussed more complex questions, such as types of data they share and with whom, gradually building an account of children's online behaviour and enabling discussion of less thought-of areas such as data harvesting and profiling A particular focus was placed on listening to what children had to say, treating them as experts on their own technology use but also attending to their struggles and concerns (Livingstone et al., 2019).

Conducted during 2019–20, the studies include 690 children living in high- and low-income countries, ranging from marginalised children living on the street to children from well-resourced, upper-middle-class families. In East Asia (Cambodia,

Indonesia, Malaysia and Thailand), we interviewed 34 focus groups with 301 social media users aged 11–19 as part of a broader study about digital technology with children with disabilities, street children, refugees, juvenile offenders and survivors of sex trafficking (Bulger and Burton, 2020). In the UK, 28 focus group interviews were conducted with 169 11- to 16-year-olds around the country to explore their understanding of interpersonal, institutional and commercial privacy in digital contexts (Stoilova et al., 2019). In Austria, in the Vienna metropolitan area, we talked with 18 focus groups of 116 children aged 8–9, 11–12 and 12–16, drawn from schools that represented the ethnic and socioeconomic make-up of the city (work in progress). The study in Belgium (in Ghent, Flanders) included 16 co-design workshops with 83 children aged 9–10 and 11–12 in two public primary schools (work in progress). In the USA, we interviewed 15 sets of parents and children (41 in total) in an upper-middle class community in Colorado (work in progress).

Children's privacy off- and online: meaningful interpersonally and contextually

It is to be expected, perhaps, that abstract notions of privacy cannot be readily deciphered by children of all ages. Yet even young children express a clear understanding of privacy as important – a matter of controlling access to oneself, of the interior world, of dignity (Laufer and Wolfe, 1977). As theorised in relation to both digital and non-digital communication contexts, privacy is "neither a right to secrecy nor a right to control but a right to *appropriate* flow of personal information" (Nissenbaum, 2010: 127). According to Gavison (1980: 423), the reasons for which we claim privacy in different situations are similar. They are related to the functions privacy has in our lives:

> the promotion of liberty, autonomy, selfhood, and human relations and furthering the existence of a free society.

In our research, we heard how privacy for children means, first and foremost, "somewhere where no one else is allowed in", "something that I don't want to show to someone" and "something that I only know". Privacy is also not being seen when taking a shower, "things that one does not even show to siblings" and "when one wants to do something by themselves", as we heard from 10-year-old boys and girls from Austrian schools. Although privacy in physical contexts is easiest to grasp, these days privacy is also a passcode for a phone; indeed, a child in Austria said, "privacy is your mobile phone".

When we say that privacy is contextually meaningful, this is to emphasise the user's perception of the context in which they share information about themselves, and the actors involved. Children are primarily conscious of their privacy in interpersonal contexts – in relation to other people, whether parents, peers or strangers. We heard from children in all the countries researched that, for example, it depended on the audience that would have access to them. Particularly problematic was whether they looked strange or silly in photos visible to their contacts when shared by their parents or school. Sharing passwords among good friends is commonly regarded as a sign of interpersonal trust. Further, children find it acceptable to share their home addresses or location with their friends but not with strangers, being highly aware of "people with bad intentions" who could abuse such information.

Privacy violations in interpersonal contexts can be intensely felt. An 11-year-old girl from Austria gave us an example: "like, that someone just takes my mobile phone and can read all my messages; unless he asks and I say he is allowed, something like that I think is shit". Similarly, when children in Flanders were asked what kind of information they share online, they would first inquire whether the researchers meant sharing with friends or with strangers. Protecting oneself from the potential abuse of social media information by strangers extends not only to personal protection but also family and home. A 13-year-old girl from Austria said that posting on social media should be done only "after a lot of thought; for example, my mum doesn't post any holiday photos when we are on holidays, like on Facebook. But when we are back, she makes a diary or something like that". Similarly, an 11-year-old girl in Flanders and a 14-year-old girl in the UK noted that they had heard of instances where social media posts about a child's family being on holiday could lead to house break-ins.

When it comes to other contexts in which personal information is shared, notably institutional contexts such as the school, and commercial contexts such as marketing and advertising, children's understanding is generally less developed, and their efforts to extend what they first learned about privacy in interpersonal contexts to these other contexts can even impede their recognition of the different logics at work (Livingstone, 2020; Stoilova et al., 2020). Children generally made generous assumptions of "good faith" regarding the decisions that their school makes about their data. Behind those expressions of trust in both schools and third party platforms was the recognition that children can no longer participate in school without extensive data processing, given that platforms are increasingly embedded in their everyday interactions with the school, for example to track their absences or use learning technologies. Hence children had little choice but to assume that only teachers would have access to their data and trusted the school not to do anything "creepy" with it, as "they're my school, they're going to keep my data safe" (boy, 11, UK). In the USA, one child observed that while he used many Google products for school, he assumed that Google would not sell his data or use it beyond "targeted advertising or impersonal data collection".

Children generally had a particularly limited understanding of the collection and processing of their personal data by commercial actors. For example, children know that Google (or the "boss of Gmail") "sees" their personal data (such as their names and passwords) (boy, 12, Flanders) or that "Google knows my favourite music, point blank" (boy, 17, USA). Among children in East Asia, "Google knows everything about us" was a frequent refrain. Children in the UK explained that "Mark Zuckerberg, he's always watching" and, aware of the Cambridge Analytica scandal, some knew that "Facebook sold the information of their users to a different company", and that "even if it is a big company, you can't always trust them". Yet children were also puzzled, believing their data to be of little value since they were still young. As children in the UK put it, "Why would somebody want to track me?" and "I just don't think that what the ordinary everyday person does on the internet is really that interesting to companies".

Digital literacy and the opaque data ecology

Does children's understanding matter? The digital environment poses citizens with multiple decisions about how to manage the flow of information about them across often-opaque contexts; here, the concept of decisional privacy is helpful, as it refers to the decisions and choices of a person with regard to their personal actions (Sarikakis and Winter, 2017). Yet research in OECD countries found that "most educational systems explicitly teach operational, critical informational, social and creative skills in primary and secondary school" but with more focus on basic operational skills and much less on programming/coding and computational thinking (OECD, 2019) and, we might add, with much less focus on gaining a critical understanding of the digital ecology.

We found that children are aware of some of the different audiences that can access their information, and they try to manage those audiences by adjusting settings or maintaining different accounts. They told us that they rely on a wide variety of privacy management techniques, including deleting emails with password information, using different passwords for games, changing passwords, using two-factor authentication (through email and phone messages) and providing false information (such as names, addresses and locations). But these decisions mostly concern individuals they are aware of (friends, family, peers or strangers), rather than the (commercial) platforms they interact with or that interact with them. The fact that social media users are offered and can change their personal privacy settings, at the interface level, may seem to afford options for active privacy protection. But even as they enable decisional privacy regarding interpersonal relations, end-user "privacy settings" obscure the fact that users are afforded little or no decisional privacy regarding the commercial data ecology invested in by platforms (Heyman et al., 2014; Livingstone, 2020). The architectures of data exploitation lack transparency and are hence difficult to scrutinise, even by the authorities that are meant to supervise the activities of commercial actors (Katyal, 2019). The opacity of commercial data sharing reveals the limits of both privacy and digital literacy. For example, children may say that "the great thing is, when you close the app [Snapchat] your images are gone" (girl, 13, Austria), and Snapchat may even tell them this. But it is not true. Snapchat maintains those images on its servers – they may no longer be available to the child's contacts, but they are still integral to the data ecology that fuels the internet.

In sum, while children have varying understandings of privacy and how their data is processed, these are not necessarily aligned with definitions deployed in regulation. It is no wonder that they do not always understand what 'data privacy' or 'data protection' means. Some talked of codes and anti-virus programmes, or ways of keeping files secret. Some confused it with Internet safety or other forms of protection, for example the use of age restrictions for films. Crucially, children mostly do not understand the underlying commercial data processing practices behind the services they routinely use. Is this a realistic expectation from children, given that such digital literacy is inaccessible even to the average adult (Park, 2013). It seems that children's expectation of agency in interpersonal contexts, and their tendency to trust familiar institutions such as their school, can be misplaced in relation to commercial contexts, resulting in a mix of frustration, misapprehension and risk. We saw them trying to piece together snippets of information, but without this necessarily adding up to the deeper knowledge they need to make wise privacy decisions. For instance, older children especially were aware that others can seek information about them: "they Google you" (boy 14, Austria). In East Asia, a group of young teens explained that even when they provided false information about their age or location, apps "already know the truth, they just let you fake it". They are increasingly aware of the Faustian deal whereby children must give their data to get access to services "for free", which, after all, they want and need, lacking funds to pay for subscription services. Hence, when it comes to 'cookies', they "click them away" by accepting them, because they are "annoying" and because one "cannot go further" without accepting them: "If it's something you're going to use and then you have to accept it, I think" (girl, 11, UK). Those children who are aware of cookies also know that they collect information about online behaviour, sometimes pointing to the advertisements for online shopping that they receive as a result.

Children's privacy protection is thus not only limited by their digital literacy but also crucial is the design and management of digital services. They are well aware that they are expected to consent to the terms and conditions imposed by platforms. This requirement is one of the most 'visible' to the user, operating directly on the user interface, which means that children are very likely to come across it. Given this awareness of terms and conditions, it is particularly problematic for their decision-making about data that users have no option but to consent if they want to participate. In interviews across the countries children expressed their scepticism regarding requests to consent to privacy policies online. Not only do they rarely study them, but they also find them too long and complex, wondering whether anyone actually reads them before agreeing or consenting to the processing of their personal data, which they saw as unavoidable. As children in Flanders put it, "they are written in such a way that no one wants to read them". Older children in Austria were also critical: "I don't believe anyone reads the terms and conditions and decides no, I won't download this app". And, of course, they are right – again, the design decisions of digital providers (and the failure of regulators to intervene on the side of the user) precludes the possibility of children's understanding or trust, both prerequisites for informed consent. Hence children learn that uninformed consent is the acceptable norm. Children in Flanders even questioned why they are presented with a choice in the first place, since they do not have a real option when asked for their agreement to the terms and conditions and/or consent for data processing or cookies, and can hardly choose "to press no" if they want or need to use the service, whether because their friends all use the service or because their school does not permit them to opt out. Children in Malaysia and the USA had similar complaints: "Nowadays, almost all the applications want you to allow access to your GPS and your media", and, "You're asking for too much information, and I just want to access these features".

Equally salient to children are the minimum age restrictions set by many online services that require users to be at least 13 years old, although children are less likely to realise that this is a result of companies seeking to comply with COPPA and/or the GDPR without having to invest in measures for younger children. But children sometimes they have also learned to create "workarounds" to bypass the barriers – for instance, that it is commonplace to "lie" about one's age, and that there are generally no adverse consequences for doing so, while there is definitely an adverse consequence to telling the truth, namely, not being able to access the chosen service. Therefore, even though some expressed concerns about "lying", young children frequently provide an older age to have access, circumventing the generally weak age assurance mechanisms operated by platforms; and sometimes they pose as adults (for example, by entering the details of a parent), which can put them at risk (Livingstone et al., 2013). According to a 12-year-old boy in Austria: "I think it's actually okay that younger people register with older ages. I find it a little stupid that social media have now changed the rules to 16 although there is nothing nasty on their sites". Such a comment reveals a confusion shared by many adults – the recent age change in WhatsApp, for example, was to avoid GDPR obligations by simply excluding children from the service rather than by taking the trouble to provide for their needs. In short, the change resulted from both regulators and businesses failing to anticipate how the introduction of data protection rules could negatively affect children's rights, but it was commonly (mis)interpreted as an internet safety measure, thereby undermining public trust in regulation.

Similarities and differences across countries

Our findings reveal some interesting contrasts because of variation in regimes of regulation and education as well as the practicalities (and inequalities) of internet access. These differences shape how children approach and understand privacy in the digital age, as well as how their data are treated by industry and regulators. Children's expectations of agency and control vary in consequence – how much they expect to be consulted, and to have their rights respected. Our findings lead us to emphasise the importance of contextualising questions of privacy within diverse experiences of childhood, including how privacy is conceptualised and enacted by children and caregivers in the datafied world.

For example, in the USA, a country that has not ratified the UNCRC because it is held to interfere with family life (US Congress, 2011), the findings reveal how greatly children want more privacy from their parents. In Austria, where children are aware of the recently introduced GDPR, the regulatory shift has afforded them a language for privacy and data rights as well as a degree of confidence in its implementation. In Flanders, children showed great interest in the fact that they have rights under the GDPR but did not testify to knowing which rights they have, let alone to exercising them. In the UK, the site of the Cambridge Analytica scandal, and with an assertive EdTech strategy (DfE, 2019), children are torn between wanting their schools to teach them digital literacy yet distrusting the companies that increasingly provide the educational infrastructure in what is commonly dubbed either a Google or a Microsoft school. In East Asia, we interviewed children who had experienced very difficult circumstances. For children living in poverty, basic literacy and tech access were the primary focus and privacy was largely beyond their sphere of attention. Notably, among those who had been exploited, some neither sought out nor expected privacy, and some considered privacy almost a risk itself – equating it with the secrecy that left them open to exploitation.

Despite these fundamental differences deriving from the structures and lifeworlds of children's lives, the value of privacy as a person's sense of control of their 'own' digital space and personal interiority remained intact across all groups. Even among the particular groups we interviewed in East Asia, where the mobile phone was often the means of their exploitation, it was also the route to access help and support. Considering the intense experiences linked to the phone, these young people talked of chatting on their phone or protecting their passwords in much the same way as those living in much more affluent or safer conditions in Europe or the USA. Despite the ways in which they had been failed by the system, they still had expectations that it would act in their best interests, including educating them as needed and protecting them and their data from further exploitation. For example, children in Cambodia expected companies to use their data to provide educational content and promote wellbeing, while a few of their peers in Thailand asked for apps that could mine their chats to identify harmful content before they could see the messages. On the other hand, the potential invasion of privacy linked to efforts precisely to protect children, especially by collecting personal information from those who are already vulnerable, is proving controversial in regulation debates, as the current debates on age assurance demonstrate (5Rights, 2021).

The similarities across countries are nothing less than striking. In all countries where we researched, children, much like the adults, around them struggled to grasp the consequences of growing up with such an omniscient and omnipresent technology. Like Altman (1975), we found that interpersonal privacy management is a universal phenomenon. Children primarily conceptualise privacy in relation to interpersonal contexts, conceiving of personal information as something they have agency and control over while 'data privacy' or 'data protection' was not something they knew much about. At the same time, we also found children to be engaged by and keen to understand the data ecology in which their lives are now enmeshed.

Wherever we interviewed children, we heard their rising tide of concern about the assaults on their privacy in a datafied world and found some misapprehensions about how personal data is collected, inferred and used by organisations, be these public institutions such as their school or, especially, businesses. Everywhere, too, we heard the outraged cry that "the internet knows everything about us" and "we have little choice", although some were more fatalistic and others more tactical in their responses. So even though we have noted the hints of contextual variation in children's understanding, far more compelling were their unified concerns, tying together expectations of education and regulation in a growing awareness of the digital environment and their relative lack of agency to manage it.

Of course, in every country children would benefit from greater educational efforts that treat them as agents, able to construct norms regarding appropriate data usage and information flows (Pangrazio and Selwyn, 2019). But the very complexity of the current digital ecology makes it hard, if not impossible, for the public to anticipate the long-term consequences of their digital footprints. Children face particular difficulties because of their age and maturity, because they may use digital products and services that are neither age-appropriate nor respectful of their rights, and because the adults responsible for them (parents, teachers and wider society) often fail to grasp or defend their interests, even undermining them through everyday surveillance practices (Leaver, 2015). Educators, too, face significant challenges regarding the effort to explain complex socio-technological changes, including scaffolding children's critical grasp of the data ecology and its algorithmic processing, technological architecture, business models, regulatory frameworks and available options for remedy or complaint. Crucial questions for educators concern their capacity (in terms of training and resources) to go beyond the usual and already-taxing expectation for a basic understanding of digital literacy to produce a dedicated and sophisticated media literacy curriculum provision that answers the questions raised by children interviewed in our research (Stoilova et al., 2020).

Children's right to be heard on matters that affect them

In many of our focus group discussions, children not only revealed their understanding, but they also expressed concerns, including their political claims and demands for change, even though these were not always articulated as such. Over and again, the discussions with children pointed to a gap between what the law requires and what happens in practice. For European children, for example, although the GDPR emphasises user rights, transparency and control, they do not feel that they are informed, and want greater control. Not only educators, but also data controllers, should make greater efforts to consider children and their needs seriously, and they should be held accountable for this by national data protection authorities. This is reinforced in the UN Committee on the Rights of the Child's General Comment No. 25 (2021), which states that businesses should be required to "maintain high standards of transparency and accountability and [...] take measures to innovate in the best interests of the child" (para. 39), along with many requirements regarding the protection of children's privacy and data in relation to the digital environment.

In the spirit of Article 12 of the UNCRC – children's right to be heard in all matters that affect them – we note here some of the demands from children that we heard during our research across the different countries.

From a 15-year-old in Austria, a request for basic respect of privacy:

> It always springs to my mind that in any case we are spied upon. I mean, alone by our mobile phones. Why? They shouldn't do it.

From a 12-year-old in the USA, a request for transparency:

> Why do you need my location? If they could just say what they're gonna do with my location, then I'd be fine.

From a 16-year-old in Thailand, a request for limiting data shared with third parties:

> Don't share every bit of information to third party companies.

From an 11-year-old in Flanders, a request not to collect location data when this is not necessary for the app or game:

> Not all apps should know my location. Because why do they need that? Why do they ask for it? Then I think to myself: 'No, this is just a simple little game'.

From a 13-year-old in the UK, a request to choose what information is collected and with whom it is shared:

> Your information is specifically yours. Like your full name, mental health, that's to do with you. So you should be able to choose who knows and who doesn't.

From a 16-year-old in the USA, a request for meaningful consent:

> You should always have consent and boundaries... consent with data, consent to have a hug. You're saying, 'These are my boundaries'.

From a 15-year-old refugee in Malaysia, a request for social media apps not to directly approach children:

> You can use it, but the parents need to be involved.

From an 11-year-old in the UK, a request for contacts to be private:

> Companies shouldn't really be poking through your contacts. Because there might be some sensitive information in there which they shouldn't get a hold of.

Clearly, children are able to reflect on their experiences and express ideas, suggestions and recommendations on ways to improve how their personal data are processed. For example, instead of plain, long and complicated pieces of text, children in Flanders suggested that details of data collection and use could be presented to them in the form of posters, mind maps, (animated) videos, online and offline games, quizzes, vlogs, sketches, explanatory pictures or drawings, letters, funny advertisements, puzzles and websites. The guidance provided by the EU's Article 29 Working Party (2018) and the UK Information Commissioner's Office (ICO, 2020) corresponds with children's own recommendations for child-specific information provision measures, such as cartoons, infographics, flowcharts, comics, animation, graphics, privacy dashboards, symbols, video and audio content, as well as gamified and interactive content.

In addition to specific formats, children consider links to their day-to-day reality very important when referring, for instance, to their school life and social media environment. Some children from the UK wanted to have conversations with their parents and teachers about the data collection and options for protection. Children in Flanders also wanted to see information about the processing of their personal information in ways that were tailored to their specific (age) group as much as possible, and wanted to be involved in its creation (for example, to feature in the information videos created for them and their peers). Whatever the presentation format, a 12-year-old boy in the USA summed up many children's view that "if people know that their information is not kept private on the internet then they might do something about it… I think it's less about not caring and more about not knowing".

Conclusions: the interdependencies of education and regulation

Communication technologies are becoming continuously more sophisticated, artificially intelligent, globally networked and commercially profitable. Protecting children's privacy in a digital world compounds the already-familiar challenges of implementing a universal rights framework, not least because children live in highly diverse contexts and political cultures around the world (UNICEF, 2014). Government actions regarding the digital environment often favour a limited vision of what 'adequate' education might entail as regards digital literacy, even in more affluent countries that could afford systematic approaches to protecting privacy, including through the introduction and implementation of data protection regulation and other forms of privacy legislation to ensure that public and private sector organisations protect children's privacy and data, as well as by promoting children's knowledge of their rights.

More fundamentally, this chapter has argued against the common view that digital literacy education and data protection regulation are mutual alternatives, to be traded off against each other in public debates as a matter of political expediency. Supposedly, if only children could understand the data ecology, they could take responsibility for their personal data and exercise wise judgement regarding the ways in which it is processed. Conveniently, since children constitute a group commonly deemed too immature to be consulted as a stakeholder, making this assumption has the effect of relieving regulators of the requirement to address children's specific vulnerabilities. The brunt of the burden to 'provide' digital literacy is instead outsourced to the educational systems, especially the teachers and, indirectly, the parents 'responsible' for their children's upbringing. In short, notwithstanding the immense socioeconomic and resource inequalities within and across countries, it is held to be the individual's responsibility to obtain, maintain and develop digital literacy skills. Yet this is a regulatory response which fails to serve emerging citizens.

By rights, literacy is a matter of civic empowerment and participation meant to operate side by side with other provisions, such as public services, public interest journalism, corporate social responsibility to name some. Yet too often, policy debates only seek to redress the damage already inflicted by private and public bodies on citizens' privacy by burdening the individual with their own protection. If regulation operated effectively in children's best interests, then the role of educators would not be limited to warning children against the potentially abusive actions of public and private organisations, for these would offer children meaningful choices that respect their evolving capacities and needs. Nor would they be called upon merely to pass on information but could be freed to creating knowledgeable participatory pathways for children to intervene in the technologies they use. But regulation relies on an effective and interventionist state, and that can in itself have problematic consequences depending on the kind of government in power; partly in consequence, there are many commercial and political interests that lobby to keep regulation weak.

Even in the wealthy countries studied, not only do disparities among educational institutions within the same country create further digital inequalities in educational environments, but also even the best available education is proving insufficient to support children's agency in a datafied world. Meanwhile in the poorer countries, the lack of resources obviates any choice between regulation and education. Moreover, the polarisation of educational versus regulatory approaches distracts from assigning responsibility to all actors involved each according to their negotiating power. The discursive construction of education *versus* regulation or even education *as* regulation does not work either in principle or in practice. In principle, children have the right to know and exercise their rights, however benevolent the state or effective its instruments, while education about legal rights is an important element of legal empowerment. In practice, education and regulation have co-evolved, with education tasked with teaching children about the available forms of governance, and with regulation embedding assumptions about what the average citizen may be expected to understand. Across a range of social issues, legal empowerment initiatives which educate the public about the law and how it impacts them have been found to increase the agency of participants both in terms of willingness to act and actual action (Goodwin and Maru, 2014). Rather than focus on either educational or regulatory dimensions of children's understanding of their privacy in a digital world, our discussion has shown that digital literacy, crucial though it is, cannot be a match for multi-country, multi-billion, oligopolistic companies that control digital platforms. It has also shown the interdependencies, by exploring how children understand both the digital environment and also its regulation, before highlighting the changes children call for from educators and regulators alike.

When it comes to ensuring that children's rights in the digital environment are truly respected, protected and fulfilled, investment of appropriate resources in both educational institutions and regulatory authorities is required. But to task educators with the responsibility of mitigating harms due in large part to the structural problem of unfair or exploitative data processing is surely to misdirect scarce educational resources. While children have the right to be informed in an honest and age-appropriate way about the nature and consequences of the processing of their personal data, it is equally or perhaps more important to implement regulation that imposes and enforces requirements for data processing in children's best interests. After all, children are hardly expected to educate themselves about other complex fields akin to data processing such as financial services or pharmaceutical drugs, in place of government regulation. Hence this chapter concludes by calling for a rights-respecting architecture for digital platforms that ensures that genuine choices are available to users, who should be fully informed about these choices, as well as preventing privacy violations and commercial exploitation of children around the world. To advance in this direction, it is imperative to give children a voice to guarantee that their best interests are taken into account in shaping and evaluating both educational and regulatory initiatives, and to recognise their interdependencies.

References

5Rights (2021) *But how do they know it is a child? Age Assurance in the Digital World.* London: 5Rights. Available at: https://5rightsfoundation.com/uploads/But_How_Do_They_Know_It_is_a_Child.pdf

Altman, I. (1975) *The Environment and Social Behavior: Privacy, Personal Space, Territory, Crowding*, Monterey, CA: Brooks/Cole.

Article 29 Data Protection Working Party, (2018) 'Guidelines on Transparency under Regulation 2016/679'. *WP260 Rev.01.* Available at: http://ec.europa.eu/newsroom/article29/item-detail.cfm?item_id=622227

Aviram, A., Gurion, B. and Eshet-Alkalai, Y. (2006) 'Towards a theory of digital literacy: Three scenarios for the next steps', *European Journal of Open, Distance and E-Learning*. Available at: https://old.eurodl.org/materials/contrib/2006/Aharon_Aviram.pdf

Buckingham, D. (2007) 'Digital media literacies: Rethinking media education in the age of the internet', *Research in Comparative and International Education*, *2*(1), 43–55. doi:10.2304/rcie.2007.2.1.43

Bulger, M. and Burton, P. (2020) *Our Lives Online: Use of Social Media by Children and Adolescents in East Asia – Opportunities, Risks and Harms.* Bangkok: UNICEF East Asia and Pacific Regional Office. Available at: www.unicef.org/eap/reports/our-lives-online

Carretero, S., Vuorikari, R. and Punie, Y. (2017) *DigComp 2.1: The Digital Competence Framework for Citizens with Eight Proficiency Levels and Examples of Use*, EUR 28558 EN. Available at: https://op.europa.eu/en/publication-detail/-/publication/3c5e7879-308f-11e7-9412-01aa75ed71a1/language-en

Chakravartty, P. and Sarikakis, K. (2006) *Media Policy and Globalisation*, Edinburgh: Edinburgh University Press.

Charter of Fundamental Rights of the European Union. OJ C 326. 2012. Available at: https://eur-lex.europa.eu/legal-content/EN/TXT/?uri=celex%3A12012P%2FTXT

De Wolf, R. and Vanden Abeele, M. (2020) 'Children's voices on privacy management and data responsibilization', *Media and Communication, 8*(4), 158–162. doi:10.17645/mac.v8i4.3722

DfE (Department for Education) (2019) 'EdTech Strategy marks "new era" for schools', Press release, 3 April. Available at: www.gov.uk/government/news/edtech-strategy-marks-new-era-for-schools

European Union (Parliament and Council) (2016) Regulation (EU) 2016/679 of 27 April 2016 on the protection of natural persons with regard to the processing of personal data and on the free movement of such data, and repealing Directive 95/46/EC (General Data Protection Regulation), *OJ* 4 May 2016, *L*119/1.

Fulton, K. (1997) *Learning in a Digital Age: Insights into the Issues*, A publication of the Milken Exchange on Information Technology, Milken Family Foundation.

Gavison, R (1980) 'Privacy and the Limits of Law', *Yale Law Journal, 89*(3). Available at: https://digitalcommons.law.yale.edu/ylj/vol89/iss3/1

Goodwin, G., and Maru, V. (2014), *What Do We Know About Legal Empowerment? Mapping the Evidence*. Namati. Available at: https://namati.org/wp-content/uploads/2014/05/Evidence-Review2.pdf

Heyman, R., De Wolf, R. and Pierson, J. (2014) 'Evaluating social media privacy settings for personal and advertising purposes', *The Journal of Policy, Regulation and Strategy for Telecommunications, Information and Media, 16*(4), 18–32. doi:10.1108/info-01-2014-0004.

ICO (Information Commissioner's Office) (2020) *Age Appropriate Design: A Code of Practice for Online Services (the Final Version)*. Available at: https://ico.org.uk/for-organisations/guide-to-data-protection/key-data-protection-themes/age-appropriate-design-a-code-of-practice-for-online-services

Katyal, S.K. (2019) 'Artificial intelligence, advertising, and disinformation', *Advertising & Society Quarterly, 20*(4). doi:10.1353/asr.2019.0026

Lansdown, G. (2014) '25 years of UNCRC: Lessons learned in children's participation', *The Canadian Journal of Children's Rights, 1*(1), 172–190. doi:10.22215/cjcr.v1i1.12

Laufer, R. and Wolfe, M. (1977) 'Privacy as a concept and a social issue: A multidimensional developmental theory', *Journal of Social Issues, 33*(3), 22–42. doi:10.1111/j.1540-4560.1977.tb01880.x

Leaver, T. (2015) 'Born Digital? Presence, Privacy, and Intimate Surveillance', in H. John and W. Qu (eds) *Reorientation: Translingual Transcultural Transmedia – Studies in Narrative, Language, Identity, and Knowledge* (pp.149–160), Shanghai: Fudan University Press.

Livingstone, S. (2020) '"It's None of Their Business!" Children's Understanding of Privacy in the Platform Society', in B. Kidron (ed.) *Freedom Security Privacy: The Future of Childhood in the Digital World* (pp.126–133), London: 5Rights Foundation. Available at https://freedomreport.5rightsfoundation.com/its-none-of-their-business-childrens-understanding-of-privacy-in-the-platform-society

Livingstone, S., Ólafsson, K. and Staksrud, E. (2013) 'Risky social networking practices among "underage" users: Lessons for evidence-based policy', *Journal of Computer-Mediated Communication, 18*(3), 303–320. doi:10.1111/jcc4.12012.

Livingstone, S., Stoilova, M. and Nandagiri, R. (2019) *Talking to Children about Data and Privacy Online: Research Methodology*, London: London School of Economics and Political Science. Available at: www.lse.ac.uk/media-and-communications/assets/documents/research/projects/childrens-privacy-online/Talking-to-children-about-data-and-privacy-online-methodology-final.pdf

Livingstone, S., van Couvering, E. and Thumim, N. (2008) 'Converging Traditions of Research on Media and Information Literacies: Disciplinary, Critical and Methodological Issues', in J. Coiro, M. Knobel, C. Lankshear and D.J. Leu (eds) *Handbook of Research on New Literacies* (pp.103–132), Mahwah, NJ: Lawrence Erlbaum Associates.

Milkaite, I. and Lievens, E. (2019) 'Children's rights to privacy and data protection around the world: Challenges in the digital realm', *European Journal of Law and Technology, 10*(1). Available at: https://ejlt.org/index.php/ejlt/article/view/674

Milkaite, I. and Lievens, E. (2020) 'Child-friendly transparency of data processing in the EU: From legal requirements to platform policies', *Journal of Children and Media, 14*(1), 5–21. doi:10.1080/17482798.2019.1701055

Nascimbeni, F. and Vosloo, S. (2019) *Digital Literacy for Children: Exploring Definitions and Frameworks*. Scoping Paper No. 01. New York: UNICEF. Available at: www.unicef.org/globalinsight/media/1271/file/%20UNICEF-Global-Insight-digital-literacy-scoping-paper-2020.pdf

Nissenbaum, H. (2010) *Privacy in Context. Technology, Policy, and the Integrity of Social Life*. Stanford, CA: Stanford University Press.

OECD (2019), "Fostering digital literacy and well-being", in Burns, T. and F. Gottschalk (eds.), *Educating 21st Century Children: Emotional Well-being in the Digital Age*, Paris: OECD Publishing. doi:10.1787/23ac808e-en

Pangrazio, L. and Selwyn, N. (2019) '"Personal data literacies": A critical literacies approach to enhancing understandings of personal digital data', *New Media & Society, 21*(2), 419–437.

Park, Yong Jin. (2013). Digital Literacy and Privacy Behavior Online. *Communication Research, 40*, 215–236.

Raffaghelli, J.E. (2020) 'Is data literacy a catalyst of social justice? A response from nine data literacy initiatives in higher education', *Education Sciences, 10*, 233–253. doi:10.3390/educsci10090233

Sarikakis, K. and Winter, L. (2017) 'Social media users' legal consciousness about privacy', *Social Media + Society*, January–March, 1–14. Available at: doi:10.1177/2056305117695325

Stoilova, M., Livingstone, S. and Nandagiri, R. (2019) *Children's Data and Privacy Online: Growing Up in a Digital Age*, London: London School of Economics and Political Science. Available at: www.lse.ac.uk/my-privacy-uk/Assets/Documents/Childrens-data-and-privacy-online-report-for-web.pdf

Stoilova, M., Livingstone, S. and Nandagiri, R (2020) 'Digital by default: Children's capacity to understand and manage online data and privacy', *Media and Communication, 8*(4). Available at: www.cogitatiopress.com/mediaandcommunication/article/view/3407

UN (United Nations) (1989) *United Nations Convention on the Rights of the Child*, Geneva. Available at: https://www.ohchr.org/en/professionalinterest/pages/crc.aspx

UN (United Nations) Committee on the Rights of the Child (2021) *General Comment No. 25 (2021) on Children's Rights in Relation to the Digital Environment*, CRC/C/GC/25. Available at https://undocs.org/CRC/C/GC/25

UNICEF (United Nations Children's Fund) (2014) *25 Years of the Convention of the Rights of the Child – Is the World a Better Place for Children?* New York: UNICEF.

US Congress (2011) *S.Res.99 – 112th Congress (2011–2012)*. Available at www.congress.gov/bill/112th-congress/senate-resolution/99

van Dijk, J.V. and Hacker, K. (2003) 'The digital divide as a complex and dynamic phenomenon', *The Information Society, 19*, 315–326. doi:10.1080/01972240309487

12

CONCLUSION

Learning to live better with data

Julian Sefton-Green and Luci Pangrazio

This book has brought together a range of perspectives and empirical accounts from around the world. We first proposed this volume prior to COVID-19 and while the pandemic of 2020/2021 has perhaps shaken understandings of the role of data in everyday life this collection does offer some key reflections on the state of organised learning about data and datafication. First, the chapters bring insights into how the datafication of education is *experienced* by key stakeholders including, teachers, administrators, students and educational bureaucrats. It has also provided insight into forms of education and learning *beyond* school; specifically, how changes and processes wrought by the digitalisation of everyday life are experienced and understood by young people. Thirdly, several chapters considered the broader contexts of schooling and education policy to glean insights into what *could* be happening or what *might* happen in the future. While these trends and possibilities are not predestined, identifying and analysing them provides an opportunity to reflect on how they might be re-shaped and re-imagined in more socially just and sustainable ways *before* they are implemented. Finally, through critical analysis and reflection, the collection provides us with a clearer sense of what *should* be happening when it comes to the datafication of education. Clearly, digital data is here to stay. Rather than resisting, ignoring or even just critiquing the use of digital data, there needs to be sustained, critical discussions of how teaching and learning can be bolstered and strengthened by digital data, rather than simply hollowed out by it.

In the introduction, we noted a trend for critical research to look downwards as it were from a point of power analysing the ripple effect of high-level decisions made by national governments and supranational corporations. While we support and espouse this perspective, research needs to account for the effects of take-up, use and meaning, making from a much wider range of social actors to show how policy decisions are actually implemented in practice. In doing so, the impact of modes of surveillance and forms of datafication on young people, their families, teachers and

DOI: 10.4324/9781003136842-12

schools and schooling in general will be revealed, rather than only looking at the effects from a single privileged perspective. A recent collection entitled 'COVID-19 from the margins' (Milan et al., 2021) exemplifies this more comprehensive approach showing how forms of datafication are negotiated and resisted in a range of everyday settings by diverse populations in local contexts. In a similar way, several of the chapters in this collection offer a similar view, focusing on how datafication is experienced first-hand by people in educational settings across the world.

Several chapters sought to inquire reflexively into what is going on in schools through a range of methods from digital ethnographies through to hermeneutic circles. In Chapter 4, Selwyn, Pangrazio and Cumbo highlighted the new types of work and power dynamics that are emerging in Australian schools as a consequence of datafication. While these are interesting findings, this is unlikely to be replicated in schools in other jurisdictions, where procurement processes and education systems will differ. For example, some countries have seen the emergence of 'intermediaries' that facilitate the procurement of digital technologies (Kerssens and van Dijck, 2021). In countries like the Netherlands, digital technologies are incorporated into education systems via intermediaries, and schools do not deal directly with tech companies. However, in Australia, it is up to schools themselves to negotiate deals with tech companies. Clearly, more empirical work is required to understand the different factors that shape procurement, incorporation and use of digital technologies and software systems in schools across the world.

Also emerging from Selwyn and colleague's chapter was insight into the complexity of the 'doing data' in schools. In addition to the traditional work of teachers and educational bureaucrats (i.e., curriculum, assessment and the general operation of the school), new knowledge and new forms of labour are required to make datafied systems work *in practice*. Juliana Raffaghelli's chapter on how educators feel about and understand these new tools and contexts for work provides a nuanced response to these issues. Clearly, teachers can see the benefit and opportunities that digital data can bring to teaching and learning; however, the data literacies required must be developed together with teachers and in context. Rafaghelli's application of 'hermeneutic circles' and the development of a template that guides this process, highlights how educator understandings can be scaffolded from more basic understandings through to the more critical and creative ways of thinking about the data they work with.

While the aforementioned 'platform gaze' is a necessary and important contribution to scholarship and public debate, in Chapter 6 Earl Aguilera and Roberto de Roock contextualised platformisation, and critically examined the array of vernacular practices, tactical interventions and local interpretations that emerged in response. Focusing specifically on the Canvas LMS system, the chapter explored the procedural ideologies of the platform and how its invisible and automated processes are navigated by teachers and students. In particular, Aguilera and de Roock were able to show how the LMS system, in combination with historically situated social practices and institutional policies, further defines and delimits the roles of teacher and student. As they explained, proceduralised ideologies are 'designed into and enacted through this software' differentiating access and agency within the system and ultimately reinforcing the teachers' more powerful position.

Several chapters expanded the analytical focus to look at the broader social and political contexts that drive and shape the datafication of education. Cristobal Cobo and Pablo Rivera-Vargas's chapter focused on the effectiveness of education policy in various countries throughout Latin America in response to the rapid uptake of Zoom during the global pandemic. They concluded that lists of 'quick tips' on how to avoid risks is insufficient in dealing with the changes brought about by digitalisation and datafication. This is a sentiment that is echoed in various ways throughout this book. Like Raffaghelli, Selwyn and colleagues and Aguilera and de Roock, Cobo and Rivera-Vargas call for a deeper understanding of what it means to be online. There needs to be closer examination of data regulation, transformation of digital governance and properly funded awareness and education programmes to support teachers, students and administrators to understand and evaluate the digital enclosures in which they work, learn and study.

However, this is not an easy or straightforward task. Indeed, Rebecca Eynon's careful deliberation on the purpose of schooling in the age of datafication set an important agenda that in effect, provoked dialogue for the remainder of the collection. Eynon detailed how digital trace data is used in schools and the ways in which it compounds inequality amongst young people. She then suggested strategies to redress the situation, including: increasing the diversity of people involved in data 'work'; improving data literacy and using participatory design projects to support students to inquire into their own data. While we started this collection with the hope that the chapters would demonstrate something similar – schools and systems exploring and developing approaches to addressing the challenges of datafication – we found that the reality of actually doing this kind of work in schools with teachers and students is difficult. We discuss this finding in more detail later in this chapter.

As Kerssen and de Haan, and Aguilera and de Roock's chapters highlight, there is considerable slippage between the concepts of datafication and platformisation. Indeed, platform architectures play a fundamental role in the creation and constitution of data. While it might be easy to think that the focus on quantification, accountability and management is a consequence of platformisation, Niels Kerssens and Mariëtte de Haan's chapter shows that it is in fact the latest step in a long-standing 'evidence-based movement' in education that has begun to dominate reform internationally. As with many other social institutions, the instrumental rationale characteristic of neoliberalism was having an effect in education long before the emergence of digital data. However, this has only been intensified through these data-driven logics. The platformisation of education is therefore a neat solution to many of the organisational and performance indicators that had begun to plague school administrators. However, with it comes the datafication of all elements of schooling – including, and perhaps most significantly, young people themselves.

Two of the chapters focused on developing school-based curriculum to improve young people's critical understandings of datafication. As these chapters highlight, this is not easy work – platform processes and architectures are 'black-boxed'

(Pasquale, 2015) or hidden from users. Working in Belgium, Jérémy Grosman, Jerry Jacques and Anne-Sophie Collard's programme involved simulating or revealing these processes and the personal information used by the YouTube recommendation algorithm to shape the digital content users see. Adapting this approach for a younger audience in South Korea, Hyeon-Seon Jeong, Yeonju Oh and Amie Kim, scaffolded their students' understanding of the 'platform imaginary'. As part of a game-like simulation, both Belgium and Korean students shaped the content users saw based on personal information. Both chapters scope out the new territory for contemporary media education. Interestingly, key to this curriculum was inverting the information asymmetry (Brunton and Nissenbaum, 2015) to give students not only a sense of the power that comes with an abundance of personal information, but also the kinds of difficult decisions that need to be made in order to do something with that data. The effectiveness of the innovative, immersive approach offered by these programmes counters the didacticism of the cybersafety discourse, which has come to dominate digital learning in recent decades.

Despite the global reach of companies like Google and Facebook, these chapters highlighted the importance of tailoring the programme to the needs of the students and the local context. In a similar way, Chapter 10, by Luo Xiaoming, highlighted the limits to a universal starting point in understanding the dimensions of datafication. Conceptualised as a 'shadow', the chapter argues that datafication in China is not primarily the domain of the commercial data broker economy but is primarily a mechanism used by those in power to track and profile Chinese citizens on the grounds of national security. This completely shifts the focus and priorities of any educational intervention on datafication in China, meaning the scope for change and resistance is not theorised and realised in the same way as it is in other chapters in the book. Luo Xiaoming's chapter is an important reminder for all researchers and educators that Western perspectives on datafication are not useful the world over. As critical data scholars (Boyd and Crawford, 2012; Iliadis and Russo, 2016; Kennedy and Bates, 2017) have pointed out, while multinational tech companies may be responsible for datafication, it is essential that researchers and educators consider the specific nature of the material, political and cultural context in which datafication is taking place. Luo Xiaoming's chapter demonstrates the different starting points for considering the most appropriate and effective way to respond to the challenges of datafication.

The final chapter looked at a different kind of response to datafication – regulation. Chapter 11, by Sonia Livingstone and colleagues, builds on her advocacy for the digital rights of the child (Livingstone and Third, 2017). In the past, Livingstone has made a strong case for bringing the 1989 UN Conventions on the Rights of the Child (UN CRC) to the digital context, arguing that the internet is typically thought of as a resource for adults and is reflected as such in policy, regulation and ideology. More recently, this has led to the development of the 5Rights Foundation, which seeks to make systemic changes to the digital world to ensure by 'design and default…that children can thrive in a digital world' (5RightsFoundation, 2020).

While much of the discussion about regulating datafication has taken place in the halls of parliament and amongst the academic elite, this chapter presents the voices of key stakeholders often overlooked in the debate – children. Like Luo Xiaoming's investigation, this chapter also demonstrates the different starting points for datafication by exploring the kinds of understandings and beliefs that emerge amongst children in different parts of the world. By listening to the voices of children, the authors highlight the need for an approach which combines both education and regulation. Children and young people felt breaches of social privacy through sharing of passcodes or surreptitious screenshots keenly. As such, regulation must be complemented by robust educational programmes that explore the affective dimensions of data. In institutional contexts, children and young people's critical understandings were less developed, meaning there was an implicit trust that institutions, organisations and private companies would protect their data. This is problematic from both a data protection and a data privacy perspective. Not only is there potential for datafied systems at any level of society to be breached but mining personal data has become key to the ways in which these companies and organisations function. Regulation therefore serves as an important 'backstop' to maintaining citizen's rights.

A manifesto for educating about data

An edited collection like this pulls together practices and activities that are, if not experimental, then certainly innovative. The case studies from Belgium and South Korea, the hermeneutic circles described by Julianna Raffaghelli are all novel pedagogies and as such this might explain some of their effects. They are also accounts of experimenting with curriculum – albeit mainly in terms of human resources. We do not yet have accounts of learning about datafication in teacher preparation or professional development contexts. This all points to the difficulty of doing this kind of work. It is difficult to devise, design and develop curriculum, difficult to implement and evaluate and certainly challenging to scale up such initiatives.

On one level, all the chapters are united in a commitment to this kind of work and indeed it was not difficult for us to find examples of need – where acute concern or anxiety about the effects of living with data have already been translated into challenges for curriculum and pedagogy. However, finding cases where schools, or local authorities or even state education systems have translated such concern into practical teaching and learning was not easy. It is perhaps unsurprising that there are more examples of studies of datafication with students at university level (such as in Chapter 6, by Aguilera and de Roock) than in school. We think this is partially because of the difficulties of researching educational reform and indeed the difficulties of engaging in educational reform where so many school systems are so focused on other kinds of outcomes than the politics or ethics of digital data. There are also good reasons why, as a field, learning about datafication is challenging. It crosses a boundary between personal experience and formal knowledge and raises a set of questions about the role of a school system engaging with its future citizens in relationship to their authority and agency.

Perhaps not surprisingly, we think a collection like this only further justifies the need to study the experiences of data in context. Rather than conceptualising of this field as an intellectual abstraction – particularly the computational and engineering dimensions at play in the process – we think many of the chapters in this collection have tried to study data in context and to get their hands dirty as it were, rather than gaze from afar. The study of Victorian teachers in Chapter 4, and indeed the students in Chapters 6, 8 and 9, the voices of the children and young people in Chapter 11, are all examples of methodological approaches that combine data science and the interpretive approach of the social sciences (Lindgren, 2021). Without this kind of understanding, we believe it will be very difficult to devise educational interventions that address the social and political implications of datafication because it will be the lived experiences of data that will support progressive pedagogy and innovative curriculum.

A key principle to further develop meaningful educational perspectives on living with data will be how we continue to conceptualise the material context of datafication. As several contributors have stressed, it is precisely this challenge of reconciling data as both invisible and consequential – material and ethereal – that makes it such a difficult topic for teaching and learning. Only when data is understood as having consequences will we be able to resist, reimagine and rearticulate datafication processes. While the kind of harms – lack of privacy, surveillance and misrecognition – are self-evidently brutal and unfair we need not let only forms of exception define acceptable norms. The fact that the effects of datafication are significantly socially unequal only compounds this challenge.

At the same time, public interest in forms of extreme exception also biases the way that educational interventions are framed. In general, we think the chapters in this volume add up to an argument that we need to avoid solutionism (Morozov, 2013), or educational approaches that are parachuted in from elsewhere for 'one-off' special programmes. Indeed, it is precisely the banality of the YouTube algorithms that make the case studies from Belgium or South Korea so resonant. There are volumes which have brought together studies of oppression and resistance (Kukutai and Taylor, 2016; D'Ignazio and Klein, 2020; Milan et al., 2021). These are important collections both in terms of content but also for the methods they offer to help build everyday vernacular understandings of datafication which often through its unexceptional and quotidian nature exerts its deepest effects.

Our collection is eclectic in terms of its scope and range. By contrast, educational interventions tend to be organised around more conventional disciplinary fields. In curriculum terms it is not clear what 'subject' – to use the language of English schooling – or curriculum space we are talking about. Is data and datafication Maths, Technology, Social Sciences, Personal and Health Education or public understanding of science? The discipline of media education permeates many of the chapters – for example the case studies in Belgium and Korea. That field and that subject has not had a happy experience in many national education curricula

(Buckingham, 2019) even if it is well placed to offer a developed theory of learning that can integrate reflection on personal experience with social and political understanding, analyses of power and technology. These questions will need resolving if this topic is to become, as all the contributors to this volume, would argue, a foundation stone in any future oriented educational philosophy.

Our final argument builds on the challenge we had in bringing together the contributors to this volume. We have noted the need for integrating tactical, educational and regulatory responses to datafication. At times we know that the distinctions we have made between regulatory and educational responses might seem tendentious. It is after all difficult to conceptualise a form of regulation without understanding, and a form of education that is not in itself in principle normative – offering versions of what might be considered proscriptive and prescriptive. We do need ways to connect up different responses to datafication – i.e., educational, legal, tactical and technical so they are mutually beneficial rather than working in isolation. Because the kind of responses we have collected here are more unusual than commonplace we get the impression that there is serendipity in the ways that any one individual might experience learning about datafication. There are very few accounts of community, neighbourhood or social responses to datafication at scale so that these different levels of intervention might come across as haphazard rather than integrated. Some of this might be a consequence of the problem's novelty but it is also difficult to avoid the sense that vested interests might well be working against a more holistic and integrated perspective.

We hope that this volume serves as a stepping stone to not just a wider conversation about data and datafication, our experience of it and its effects but to provide evidence and practical guidance for families, teachers, schools, educators and other policymakers to come together to address a significant challenge facing all societies across the world. Learning to live well with data is, if it isn't already, a serious challenge for future generations.

References

5RightsFoundation. (2020). *Introduction*. Retrieved from https://5rightsfoundation.com

Boyd, D., & Crawford, K. (2012). Critical questions for big data. *Information, Communication & Society, 15*(5), 662–679.

Brunton, F., & Nissenbaum, H. (2015). *Obfuscation: A User's Guide for Privacy and Protest*. Massachusetts, CA: The MIT Press.

Buckingham, D. (2019). *The Media Education Manifesto*. London: John Wiley & Sons.

D'Ignazio, C., & Klein, L. F. (2020). *Data Feminism*. Cambridge MA: MIT Press.

Iliadis, A., & Russo, F. (2016). Critical data studies: An introduction. *Big Data & Society, Advanced Online Edition*, 3, 1–7.

Kennedy, H., & Bates, J. (2017). Data power in material contexts: Introduction. *Television & New Media, 18*(8), 701–705.

Kerssens, N., & van Dijck, J. (2021). The platformization of primary education in The Netherlands. *Learning, Media and Technology*. doi: 10.1080/17439884.2021.1876725

Kukutai, T., & Taylor, J. (Eds.). (2016). *Indigenous Data Sovereignty: Toward an agenda* (Vol. 38). Camberra: ANU Press. http://www.jstor.org/stable/j.ctt1q1crgf

Lindgren, S. (2021). *Data Theory*. Cambridge, UK: Polity.

Livingstone, S., & Third, A. (2017). Children and young people's rights in the digital age: An emerging agenda. *New Media & Society*, *19*(5), 657–670.

Milan, S., Trere, E., & Masiero, S. (2021). *COVID-19 From the Margins: Pandemic Invisibilities, Policies and Resistance in the Datafied Society*. Amsterdam, The Netherlands: Institute of Network Cultures.

Morozov, E. (2013). *To Save Everything, Click Here: Technology, Solutionism and the Urge to Solve Problems That Don't Exist*. London: Penguin.

Pasquale, F. (2015). *The Black Box Society: The Secret Algorithms that Control Money and Information*. Cambridge, Massachusetts: Harvard University Press.

13

AFTERWORD

The future of datafication in education? Clouds, bodies and ethics

Ben Williamson

Introduction

This excellent collection, *Learning to Live with Datafication: Educational Case Studies and Initiatives from Across the World*, edited by Luci Pangrazio and Julian Sefton-Green, emphasises the necessity of research that focuses on situated effects, uses and interpretations of data-driven technologies in education. It asks important questions about the experiences of living, teaching and learning in datafied systems, how datafication might evolve in the future, what should be done about it, and calls for informed critical discussion about how teaching and learning might be strengthened by datafication, as well as its present and emerging risks, and possible forms of response or resistance.

Complementing studies of datafication focused on governmental policy initiatives or commercial activity (see Grek et al., 2021; Wyatt-Smith et al., 2021), the collection prioritises attention to the practices and experiences of datafication by diverse stakeholders in local contexts. In this respect, it resonates with recent research examining how platform architectures interact with policy regimes and users in context-specific educational settings, and analytical attention to the actual enactments of datafication and platformisation in education (Decuypere et al., 2021). Yet, as the present collection amply demonstrates, there remains pressing need for research into the varied experiences and practices of datafication and platforms, as well as emerging cloud infrastructures, machine learning and AI applications in education.

In this brief afterword, I outline further areas where such research remains essential but so far underdeveloped: the experiences and practices of teaching and learning in cloud computing infrastructures; new forms of datafication which extract intimate information about learning and teaching from bodies and faces; and the

DOI: 10.4324/9781003136842-13

experiences and practices of negotiating data ethics and regulation in education. These represent various futures of datafication in education – futures which are already emerging in the present, but seem likely to exert effects on teaching and learning in coming years, and which will demand the critical attention of researchers, practitioners and policy professionals alike.

Teaching and learning in the cloud classroom

As the chapters in this volume demonstrate, datafication in education materialises in practices and experiences in various ways in diverse contexts. It challenges contemporary perceptions of the very purposes of education, as Rebecca Eynon argues in her contribution. It also reproduces a historical series of education reform efforts that Niels Kerssens and Mariëtte de Haan characterise as 'instrumental rationality', with technologies and techniques of datafication as core instruments for maximising learning effectivity and performance. New conceptions of purpose or reformatory visions associated with datafication in education are not simply external 'framings': they are encoded in the very technologies that increasingly inhabit and execute programmed commands in classrooms. For example, in Chapter 6, Earl Aguilera and Roberto de Roock argue that specific ideologies are encoded in data-processing software systems as computational 'procedures' that generate 'transactional' models of pedagogy, curriculum and assessment.

Much existing research on datafication focuses on software packages and platforms such as teacher apps, learning management systems and other student-facing platforms. However, many of these platforms also depend on global corporate cloud infrastructures, with multinational cloud operators increasingly seeking to expand across the educational sector. Major global cloud computing operators, including Amazon Web Services, Google Cloud and Microsoft Azure, host and power many of the digital services and platforms that schools, colleges and universities routinely use. For example, two of the market-dominant learning management system providers, Canvas and Blackboard, rely on AWS for server hosting, computing power and data analysis capacity. Many education technology platforms – from small-scale start-ups to global edu-business operations – also depend on corporate clouds.

According to a major study of cloud operators' expansion in higher education, universities in the US and UK in particular have become dependent on corporate cloud infrastructures for many of their core digital services and functions since around 2015:

> Instead of running IT services with on-site teams and on infrastructure owned by organisations, services are now often deployed on public cloud infrastructure. [...] However, this operational paradigm shift also leads to a change in control. While, before, user data would remain on infrastructure controlled by an organisation, this data is now stored and processed by an external operator.
>
> (Fiebig et al., 2021, p.1)

Schools, too, are increasingly 'cloudified'. The cloud-based Google Classroom platform has a reach to 150 million students worldwide, and integrates with a marketplace of hundreds of other third-party platforms (Perrotta et al., 2021). AWS, MS Azure and Google Cloud also generate 'data lakes' of huge volumes of data for analysis, by combining information from institutional systems with the full range of platforms used for pedagogic, management or assessment purposes. They also run extensive training and certification programmes for educators to become cloud experts.

These examples indicate how schools and universities alike are increasingly dependent upon corporate clouds, which carry encoded assumptions or ideologies of how education could or should operate. These clouds may then be reshaping how schools and universities function, ultimately configuring new kinds of cloud classrooms, cloud campuses and cloud educators that are interdependent with interoperable infrastructures and marketplaces of platform integrations.

Why does this matter? The cloud introduces new algorithmic capacities of automated machine learning, prediction and pre-emption that, according to Amoore (2020, p.11), are 'redefining our lives' and 'having specific and durable effects on contemporary society'. The algorithms of the cloud, Amoore argues, define what matters in the world from condensing a multiplicity of data points into simplified outputs for action, and are thus 'changing the processes by which people and things are rendered perceptible and brought to attention' (p.15). This is a significant point for the experiences and practices of teaching and learning in the cloud, raising key questions for analysis. How do Google Cloud, MS Azure and AWS function to render students perceptible from their scattered data points in the cloud? For whom (or what) are these datafied students brought to attention? How are educators trained and certified to make use of the cloud? With what effects in terms of pedagogic, curricular, assessment or management practices and interventions? Cloud computing introduces new capacities of data analysis and algorithmic intervention that may shape and steer educators' practices and students' learning experiences. Clouds may even exert control over the 'means of study' by determining what kinds of educational tasks are possible to implement (Fiebig et al., 2021, p.7). For teachers, the cloud classroom promotes a new form of professional practice that is augmented by 'automated thinking' (Sellar and Gulson, 2021). For students, the cloud classroom is a new learning space in which they are made perceptible and brought to attention for intervention in ways defined by algorithms, including 'pre-emptive' action (Witzenberger and Gulson, 2021).

The editors of this volume suggest previous research on datafication in education has tended to privilege a 'platform gaze' and instead propose privileging the experiences and practices of datafication in context. With the expansion of cloud technologies, perhaps, the context of experience and practice is itself the cloudified classroom, and its recursive capacity to intervene in pedagogic practice and learning experience. Analysis of the cloud classroom requires both a platform gaze and close attention to lived experience in order to describe and conceptualise the ways specific cloud capacities interweave with practices and experiences.

Living intimately with data bodies

Contrasted with the global operators of the cloud, datafication in education is also becoming more focused on the intimate details of the human body. New kinds of intimate analytics that can trace physical movements, facial expressions, and even brain activity, are already emerging in scientific laboratories and commercial research and development settings. Thus, while in this volume Neil Selwyn and colleagues present a 'hierarchy' of 'data classes' of experts in schools, beyond the school exists another data class of experts who can crisscross the computing, data and human sciences to produce new insights about students' learning, cognition, emotion and neural states.

The emerging field of 'learning engineering', for example, combines the learning sciences and learning analytics with developments in AI, software engineering and sensor technologies from the psychological, cognitive and neural sciences (Williamson, 2020). It has become a catchy term in several research-intensive US universities. The Chan Zuckerberg Initiative – the philanthropic and investment vehicle of Facebook founder Mark Zuckerberg – is one the most influential organisations to promote learning engineering. Its director of learning science has described it as an 'interdisciplinary effort' enabling learning engineers to gather and analyse evidence 'at multiple levels, from clickstreams, motion position data, speech streams, gaze data, biometric and brain sensing, to more abstracted feature sets from all this evidence' (Saxberg, 2018, p.viii).

Learning engineering is related to a range of other developments in learning analytics and educational data science, including the development of facial emotional detection in the subfield of 'emotion learning analytics'. McStay (2020) has referred to 'emotion AI' in the educational field as a form of 'automated industrial psychology' that utilises biosensors, facial vision algorithms and analytics to trace autonomic biological signals as proxies of emotional sensations or non-cognitive behaviours related to learning. Similarly, brain-scanning devices have been developed to track signals of cognition and attention during learning experiences, in some cases even to provide neurofeedback in the shape of attention prompts or 'neuro-personalised' adaptations to course content (Williamson, 2018). Both facial emotion detection and brain scanning in education have been the source of widespread criticism, not least where they have been deployed in schools in China. As Lio Xiaoming reports in this volume, the Chinese education system is already highly datafied, with very different approaches to student privacy and rights than many other contexts reported in this book. That said, while attention tends to fall on examples of intimate data analytics such as emotion sensing and neurotechnology in China, the underpinning approaches they deploy are thoroughly embedded in learning sciences, analytics and engineering fields in Australian, north American and European contexts too.

What are the consequences of intimate and automated forms of psychological, cognitive and neural datafication in education? As Pickup (2021) argues, systems of datafication 'invent' student identities, rendering them as 'digital proxies' or 'informational people' who are known in bits and bytes and thereby made amenable to evaluation, prediction and potential targeted intervention. With intimate data analytics, students are anatomised biologically as 'data bodies', with the autonomic biological signals traced from their faces and bodies treated as proxies of internal states of attention, cognition, neural function and affect. These identified states may then be brought to attention and targeted for intervention or improvement.

But these are quantified bodies constructed, stored and modelled from data in computers rather than subjectively embodied and experienced in context. Intimate data analytics approaches such as those espoused by learning engineering advocates seek to render the autonomic, unconscious, embodied and automatic signals that indicate learning amenable to being machine-read and analysed, and from there potentially opened up to enhancement and engineering. To date, it remains unclear how these emerging developments in intimate bodily datafication will evolve in the near future. With wealthy organisations such as the Chan Zuckerberg Initiative supporting them, it will be an urgent priority to both unpack their scientific and technical underpinnings, and the practices, experiences and effects that result for teachers and students.

Experiencing and practising data ethics, rights and regulation

Many of the chapters in this volume rightly highlight the importance of addressing data literacy, rights and ethics to ensure adequate regulation of datafication in education. Juliana Raffaghelli reports on practices to raise educators' and students' data literacy; Cristobal Cobo and Pablo Rivera-Vargas argue for stronger regulation and governance of datafied education, based on their case study of the videoconferencing platform Zoom; and the two chapters focused on the YouTube recommendation algorithm – by Jérémy Grosman, Jerry Jacques and Anne-Sophie Collard, and by Hyeon-Seon Jeong, Yeonju Oh and Amie Kim – emphasise the potential ways datafication can become the object of pedagogic interrogation. Chapter 11, by Sonia Livingstone and consortium, however, reports very limited opportunities for children to learn critical data literacies, slowly emerging digital rights frameworks and weak regulation of educational data.

Data ethics, rights and regulations represent important aspects of the future of datafication in education for two reasons. First, there are positive emerging signs that these issues are being taken seriously, including in the European Union, and are likely to play a significant part in how datafication is practised and experienced in schools. Second, however, it is also likely that data ethics, rights and regulation will remain significantly contested, unevenly implemented and subjected to political and commercial challenge. Such contests would reflect the wider context of data ethics and regulatory developments, where there remain significant concerns of 'ethics washing' through self-regulatory frameworks and industry-led 'checklists' (Greene, Hoffman and Stark, 2019).

Already, in education, various data and AI ethics frameworks have been proposed or produced, though often with the participation of the very EdTech industry that depends on datafication for the functioning and monetisation of its products. Entrepreneurial academics in the learning analytics field have begun proposing technical ways to tackle problems such as algorithmic bias and discrimination. The European Union has assembled an expert group to produce an ethical framework for the use of AI in education, while under separate AI regulatory proposals, the EU has also designated certain aspects of AI in education as 'high risk'. This all points to a complex environment of ethical and regulatory developments that will not be resolved quickly.

One important intervention in the UK context is a detailed proposal for the regulation of both the commercial and governmental uses of student data. Produced by the Digital Futures Commission for the 5Rights Foundation, it makes a series of regulatory recommendations from an explicitly child rights perspective (Day, 2021). The 5Rights report is an important intervention into attempts to regulate educational data. Yet, from the perspective of this collection, significant questions remain about how data ethics and regulation may be interpreted, experienced and enacted in diverse contexts. As the 5Rights report indicates, the burden for ensuring ethical and regulatory compliance falls often upon schools, while EdTech companies remain unaffected. This situation reveals a significant power asymmetry, with companies free to market products – often with weak evidence of pedagogic benefit – and schools expected to carry the burden of maintaining data protection and privacy due diligence.

From the research approach endorsed in this book, questions remain about how data ethics, rights and regulatory efforts may shape the practices and experiences of datafication in schools. Research on these matters might focus on teachers' understandings and enactments of their ethical obligations and practices; on the ways that different ethics frameworks and advice are disseminated to schools; and how they are translated into action in classrooms. If commercial uses of school and student data becomes the subject of regulation, what are the effects on pedagogy, curriculum and assessment? Indeed, given the wide uptake of EdTech, especially since the COVID-19 pandemic, some teachers may feel data regulation restricts practices they have come to consider part of their professional pedagogic routines. While data ethics, rights and regulation absolutely should play a part in shaping the future of datafication in education, this is likely to remain a contested, conflicted area of development and counter-development for some years.

Final thoughts

Cloud infrastructures, intimate analytics, and data ethics and regulatory frameworks will, in various ways, shape the futures of datafication in education. This book provides much-needed empirical and theoretical analysis of the ways datafication in education is enacted, practised and experienced. It remains important, however, as the editors argue in their introduction, to be attentive to emerging developments and possible futures of datafication. This is not about speculation, but about detecting signals of change, evolution, and emerging risks and acting in the present to ensure datafication in education benefits educators, learners and the public education sector as a whole. This book represents an important step in that direction.

References

Amoore, L. 2020. *Cloud Ethics: Algorithms and the attributes of ourselves and others*. London: Duke University Press.

Day, E. 2021. *Governance of data for children's learning in UK state schools*. Digital Futures Commission, 5Rights Foundation.

Decuypere, M., Grimaldi, E. and Landri, P. 2021. Introduction: Critical studies of digital education platforms. *Critical Studies in Education*, 62(1): 1–16.

Fiebig, T., Gürses, S., Gañán, C.H., Kotkamp, E., Kuipers, F., Lindorfer, M., Prisse, M. and Sari, T. 2021. Heads in the Clouds: Measuring the Implications of Universities Migrating to Public Clouds. arXiv preprint: arXiv:2104.09462.

Greene, D., Hoffman, A.L. and Stark, L. 2019. 'Better, Nicer, Clearer, Fairer: A Critical Assessment of the Movement for Ethical Artificial Intelligence and Machine Learning.' *Proceedings of the 52nd Hawaii International Conference on System Sciences*. https://scholarspace.manoa.hawaii.edu/bitstream/10125/59651/0211.pdf

Grek, S., Maroy, C. and Verger, A. (eds). 2021. *Accountability and Datafication in the Governance of Education*. London: Routledge.

McStay, A. 2020. Emotional AI and EdTech: serving the public good? *Learning, Media and Technology*, 45(3), 270–283.

Perrotta, C., Gulson, K.N., Williamson, B. & Witzenberger, K. 2021. Automation, APIs and the distributed labour of platform pedagogies in Google Classroom. *Critical Studies in Education*, 62(1), 97–113.

Pickup, A. 2021. Toward a historical ontology of the infopolitics of data-driven decision-making (DDDM) in education. *Educational Philosophy and Theory*. doi:10.1080/00131857.2021.1935232

Saxberg, B. 2018. 'Preface', In D. Niemi, R.D. Pea, B. Saxberg and R.E. Clark (eds). *Learning Analytics in Education*. Charlotte, NC, Information Age Publishing, vii–x.

Sellar, S. & Gulson, K.N. 2021. Becoming information centric: The emergence of new cognitive infrastructures in education policy. *Journal of Education Policy*, 36(3), 309–326.

Williamson, B. 2018. Brain data: scanning, scraping and sculpting the plastic learning brain through neurotechnology. *Postdigital Science & Education*, 1(1): 65–86.

Williamson, B. 2020. New digital laboratories of experimental knowledge production: Artificial intelligence and education research. *London Review of Education*, 18(2): 209–220.

Witzenberger, K. & Gulson, K.N. 2021. Why EdTech is always right: Students, data and machines in pre-emptive configurations. *Learning, Media and Technology*. doi:10.1080/17439884.2021.1913181

Wyatt-Smith, C., Lingard, B. and Heck, E. (eds). 2021. *Digital Disruption in Teaching and Testing: Assessments, Big Data, and The Transformation Of Schooling*. Abingdon: Routledge.

INDEX

Page numbers in *italics* refer to figures; **bold** refer to tables and page numbers followed by 'n' refer to notes numbers.

For Product Safety Concerns and Information please contact our EU representative GPSR@taylorandfrancis.com Taylor & Francis Verlag GmbH, Kaufingerstraße 24, 80331 München, Germany